Revise
A2

AQA
Mathematics

Peteshaw

Contents

Chapter 3 Mechanics 2

Chapter 4 Decision Mathematics 2

Specification information

AQA Mathematics

MODULE	SPECIFICATION TOPIC	CHAPTER REFERENCE	STUDIED IN CLASS	REVISED	PRACTICE QUESTIONS
Core 3 (C3)	Algebra and functions	1.2			
	Trigonometry	1.4			
	Exponentials and logarithms	1.2			
	Differentiation	1.5			
	Integration	1.6			
	Numerical methods	1.7			
Core 4 (C4)	Algebra and functions	1.1			
	Coordinate geometry in the (x, y) plane	1.3			
	Sequences and series	1.1			
	Trigonometry	1.4			
	Exponentials and logarithms	1.6			
	Differentiation and integration	1.5, 1.6			
	Vectors	1.8			
Statistics 2 (S2)	Discrete random variables	2.1			
	Poisson distribution	2.1			
	Continuous random variables	2.2			
	Estimation	2.4			
	Hypothesis testing	2.3, 2.4			
	χ^2 contingency table tests	2.4			
Mechanics 2 (M2)	Moments and centres of mass	3.3, 3.4			
	Kinematics	3.1, 3.2			
	Newton's laws of motion	3.1			
	Application of differential equations	3.1			
	Uniform circular motion	3.5			
	Work, energy and power	3.7, 3.8			
	Vertical circular motion	3.6			
Decision 2 (D2)	Critical path analysis	4.3			
	Allocation	4.5			
	Dynamic programming	4.6			
	Network flows	4.2			
	Linear programming	4.4			
	Game theory for zero-sum games	4.1			

Examination analysis

The Advanced Level Mathematics GCE consists of AS (50%) and A2 (50%). AS consists of C1 + C2 + one of M1, S1, D1. A2 consists of C3 + C4 + one of M1, S1, D1, M2, S2, D2 (following any dependency rules).

C3	A2	Scientific / graphics calculator	1 hr 30 min exam	$16\frac{2}{3}$% of A Level
C4	A2	Scientific / graphics calculator	1 hr 30 min exam	$16\frac{2}{3}$% of A Level
M1/M2	AS/A2	Scientific / graphics calculator	1 hr 30 min exam*	$16\frac{2}{3}$% of A Level
S1/S2	AS/A2	Scientific / graphics calculator	1 hr 30 min exam*	$16\frac{2}{3}$% of A Level
D1/D2	AS/A2	Scientific / graphics calculator	1 hr 30 min exam	$16\frac{2}{3}$% of A Level

* 1 hr 15 min with coursework

AS/A2 Level Mathematics courses

AS and A2

The AQA Mathematics A Level course is in two parts, with three separate modules in each part. Students first study the AS (Advanced Subsidiary) course. Some will then go on to study the second part of the A Level course, called A2. Advanced Subsidiary is assessed at the standard expected halfway through an A Level course: i.e., between GCSE and Advanced GCE. This means that the AS and A2 courses are designed so that difficulty steadily increases:

- AS Mathematics builds from GCSE Mathematics
- A2 Mathematics builds from AS Mathematics.

How will you be tested?

Assessment units

For AS Mathematics, you will be tested by three assessment units. For the full A Level in Mathematics, you will take a further three units. AS Mathematics forms 50% of the assessment weighting and A2 Mathematics forms the other 50% for the full A Level.

3 units for AS   *3 units for A2*

Most units can be taken in either January or June, but some A2 units must be taken at the end of the course. There is a lot of flexibility about when exams can be taken and the diagram below shows just some of the ways that the assessment units may be taken for AS and A Level Mathematics.

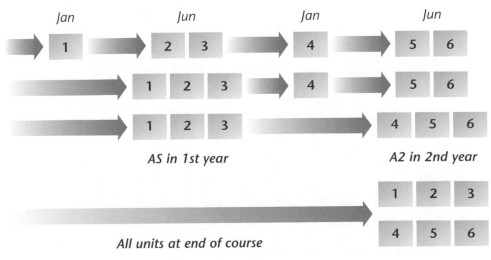

5

If you are disappointed with a module result, you can resit the module. There is no restriction on the number of times a module may be attempted. The best available results for each module will count towards the final grade.

A2 progression

After having studied AS Mathematics, to continue studying Mathematics to A Level you will need to take three further units of Mathematics, at least two of which are at A2 standard.

Synoptic assessment

Synoptic assessment in Mathematics addresses candidates' understanding of the connections between different elements of the subject. It involves the explicit drawing together of knowledge, understanding and skills learned in different parts of the A Level course, through using and applying methods developed earlier to solving problems. Making and understanding connections in this way is intrinsic to learning mathematics.

For A2 modules you will be asked to demonstrate knowledge of modules studied at AS Level. The contents of C1 and C2 are assumed for C3 and the contents of C1, C2 and C3 are assumed for C4. Similarly in the application modules, a knowledge of S1 is assumed for S2, a knowledge of M1 is assumed for M2 and a knowledge of D1 is assumed for D2.

Coursework

A coursework option is available for the first mechanics and statistics components, MM1A and MS1A. The coursework carries 25% of the marks on each of these components. However, some schools choose to enter candidates for the alternative versions of these components, MM1B and MS1B, which do not include coursework.

Your school will inform you about any coursework requirements.

Key skills

It is important that you develop your key skills throughout your AS and A2 courses. These are important skills that you need whatever you do beyond AS and A Levels. To gain the key skills qualification, which is equivalent to an AS Level, you will need to collect evidence together in a 'portfolio' to show that you have attained Level 3 in Communication, Application of number and Information technology. You will also need to take a formal test in each key skill. You will have many opportunities during AS and A2 Mathematics to develop your key skills.

It is a worthwhile qualification, as it demonstrates your ability to put your ideas across to other people, collect data and use up-to-date technology in your work.

What skills will I need?

For A Level Mathematics (AS and A2), you will be tested by assessment objectives: these are the skills and abilities that you should have acquired by studying the course. The assessment objectives for A Level Mathematics are shown below.

Candidates should be able to:

- recall, select and use their knowledge of mathematical facts, concepts and techniques in a variety of contexts

- construct rigorous mathematical arguments and proofs through use of precise statements, logical deduction and inference and by the manipulation of mathematical expressions, including the construction of extended arguments for handling substantial problems presented in unstructured form

- recall, select and use their knowledge of standard mathematical models to represent situations in the real world; recognise and understand given representations involving standard models; present and interpret results from such models in terms of the original situation, including discussion of the assumptions made and refinement of such models

- comprehend translations of common realistic contexts into mathematics; use the results of calculations to make predictions, or comment on the context; and, where appropriate, read critically and comprehend longer mathematical arguments or examples of applications

- use contemporary calculator technology and other permitted resources (such as formulae booklets or statistical tables*) accurately and efficiently; understand when not to use such technology, and its limitations; give answers to appropriate accuracy.

* You can find a copy of the formulae booklet and statistical tables in the A Level Mathematics section of the AQA website, www.aqa.org.uk

Exam technique

What are the examiners looking for?

Examiners use certain words in their instructions to let you know what they are expecting in your answer. Make sure that you know what they mean so that you can give the right response.

Write down, state

You can write your answer without having to show how it was obtained. There is nothing to prevent you doing some working if it helps you, but if you are doing a lot then you might have missed the point.

Calculate, find, determine, show, solve

Make sure that you show enough working to justify the final answer or conclusion. Marks will be available for showing a correct method.

Deduce, hence

This means that you are expected to use the given result to establish something new. You must show all of the steps in your working.

Draw

This is used to tell you to plot an accurate graph using graph paper. Take note of any instructions about the scale that must be used. You may need to read values from your graph.

Sketch

If the instruction is to sketch a graph then you don't need to plot the points but you will be expected to show its general shape and its relationship with the axes. Indicate the positions of any turning points and take particular care with any asymptotes.

Find the exact value

This instruction is usually given when the final answer involves an irrational value such as a logarithm, e, π or a surd. You will need to demonstrate that you can manipulate these quantities so don't just key everything into your calculator or you will lose marks.

If a question requires the final answer to be given to a specific level of accuracy then make sure that you do this or you might needlessly lose marks.

Some dos and don'ts

Dos

Do read the question

- Make sure that you are clear about what you are expected to do. Look for some structure in the question that may help you take the right approach.
- Read the question *again* after you have answered it as a quick check that your answer is in the expected form.

Do use diagrams

- In some questions, particularly in mechanics, a clearly labelled diagram is essential. Use a diagram whenever it may help you understand or represent the problem that you are trying to solve.

Do take care with notation

- Write clearly and use the notation accurately. Use brackets when they are required.
- Even if your final answer is wrong, you may earn some marks for a correct expression in your working.

Do learn relevant formulae

- Each module specification contains a list of formulae that *will not be given* in the examination. Make sure that you learn these formulae and practise using them.
- Some formulae *will be given* in the Examination Formulae Booklet. Make sure you know what they are and where to find them. This will help you refer to them quickly in the examination so that you don't waste time.

Do avoid silly answers

- Check that your final answer is sensible within the context of the question.

Do make good use of time

- Choose the order in which you answer the questions carefully. Do the ones that are easiest for you first.
- Set yourself a time limit for a question depending on the number of marks available.
- Be prepared to leave a difficult part of a question and return to it later if there is time.
- Towards the end of the exam make sure that you pick up all of the easy marks in any questions that you haven't got time to answer fully.

Don'ts

Don't work with rounded values

- There may be several stages in a solution that produce numerical values. Rounding errors from earlier stages may distort your final answer. One way to avoid this is to make use of your calculator memories to store values that you will need again.

Don't cross out work that may be partly correct

- It's tempting to cross out something that hasn't worked out as it should. Avoid this unless you have time to replace it with something better.

Don't write out the question

- This wastes time. The marks are for your solution!

What grade do you want?

Everyone should be able to improve their grades but you will only manage this with a lot of hard work and determination. The details given below describe a level of performance typical of candidates achieving grades A, C or E. You should find it useful to read and compare the expectations for the different levels and to give some thought to the areas where you need to improve most.

Grade A candidates

- Recall or recognise almost all the mathematical facts, concepts and techniques that are needed, and select appropriate ones to use in a variety of contexts.
- Manipulate mathematical expressions and use graphs, sketches and diagrams, all with high accuracy and skill.
- Use mathematical language correctly and proceed logically and rigorously through extended arguments or proofs.
- When confronted with unstructured problems they can often devise and implement an effective solution strategy.
- If errors are made in their calculations or logic, these are sometimes noticed and corrected.
- Recall or recognise almost all the standard models that are needed, and select appropriate ones to represent a wide variety of situations in the real world.
- Correctly refer results from calculations using the model to the original situation; they give sensible interpretations of their results in the context of the original realistic situation.
- Make intelligent comments on the modelling assumptions and possible refinements to the model.
- Comprehend or understand the meaning of almost all translations into mathematics of common realistic contexts.
- Correctly refer the results of calculations back to given context and usually make sensible comments or predictions.
- Can distil the essential mathematical information from extended pieces of prose having mathematical content.
- Comment meaningfully on the mathematical information.
- Make appropriate and efficient use of contemporary calculator technology and other permitted resources, and are aware of any limitations to their use.
- Present results to an appropriate degree of accuracy.

Grade C candidates

- Recall or recognise most of the mathematical facts, concepts and techniques that are needed, and usually select appropriate ones to use in a variety of contexts.
- Manipulate mathematical expressions and use graphs, sketches and diagrams, all with a reasonable level of accuracy and skill.
- Use mathematical language with some skill and sometimes proceed logically through extended arguments or proofs.
- When confronted with unstructured problems they sometimes devise and implement an effective and efficient solution strategy.
- Occasionally notice and correct errors in their calculations.
- Recall or recognise most of the standard models that are needed and usually select appropriate ones to represent a variety of situations in the real world.

- Often correctly refer results from calculations using the model to the original situation; they sometimes give sensible interpretations of their results in context of the original realistic situation.
- Sometimes make intelligent comments on the modelling assumptions and possible refinements to the model.
- Comprehend or understand the meaning of most translations into mathematics of common realistic contexts.
- Often correctly refer the results of calculations back to the given context and sometimes make sensible comments or predictions.
- Distil much of the essential mathematical information from extended pieces of prose having mathematical content.
- Give some useful comments on this mathematical information.
- Usually make appropriate and effective use of contemporary calculator technology and other permitted resources, and are sometimes aware of any limitations to their use.
- Usually present results to an appropriate degree of accuracy.

Grade E candidates

- Recall or recognise some of the mathematical facts, concepts and techniques that are needed, and sometimes select appropriate ones to use in some contexts.
- Manipulate mathematical expressions and use graphs, sketches and diagrams, all with some accuracy and skill.
- Sometimes use mathematical language correctly and occasionally proceed logically through extended arguments or proofs.
- Recall or recognise some of the standard models that are needed and sometimes select appropriate ones to represent a variety of situations in the real world.
- Sometimes correctly refer results from calculations using the model to the original situation; they try to interpret their results in the context of the original realistic situation.
- Sometimes comprehend or understand the meaning of translations into mathematics of common realistic contexts.
- Sometimes correctly refer the results of calculations back to the given context and attempt to give comments or predictions.
- Distil some of the essential mathematical information from extended pieces of prose having mathematical content; they attempt to comment on this mathematical information.
- Candidates often make appropriate and efficient use of contemporary calculator technology and other permitted resources.
- Often present results to an appropriate degree of accuracy.

The table below shows how your uniform standardised mark is translated.

USM	80%	70%	60%	50%	40%
grade	A	B	C	D	E

To achieve an A* grade in Mathematics, you need to achieve a grade A overall (an average of 80 or more on uniform mark scale) for the whole A Level qualification and an average of 90 or more on the uniform mark scale in Core 3 and Core 4. It is awarded for A Level qualification only and not for the AS qualification or individual units.

Four steps to successful revision

Step 1: Understand

- Study the topic to be learned slowly. Make sure you understand the logic or important concepts.
- Mark up the text if necessary – underline, highlight and make notes.
- Re-read each paragraph slowly.

GO TO STEP 2

Step 2: Summarise

- Now make your own revision note summary:
  What is the main idea, theme or concept to be learned?
  What are the main points? How does the logic develop?
  Ask questions: Why? How? What next?
- Use bullet points, mind maps, patterned notes.
- Link ideas with mnemonics, mind maps, crazy stories.
- Note the title and date of the revision notes
  (e.g. Mathematics: Differentiation, 3rd March).
- Organise your notes carefully and keep them in a file.

This is now in **short-term memory**. You will forget 80% of it if you do not go to Step 3.
GO TO STEP 3, but first take a 10 minute break.

Step 3: Memorise

- Take 25 minute learning 'bites' with 5 minute breaks.
- After each 5 minute break test yourself:
  Cover the original revision note summary
  Write down the main points
  Speak out loud (record yourself)
  Tell someone else
  Repeat many times.

The material is well on its way to **long-term memory**.
You will forget 40% if you do not do step 4. **GO TO STEP 4**

Step 4: Track/Review

- Create a Revision Diary (one A4 page per day).
- Make a revision plan for the topic, e.g. 1 day later, 1 week later, 1 month later.
- Record your revision in your Revision Diary, e.g.
  Mathematics: Differentiation, 3rd March 25 minutes
  Mathematics: Differentiation, 5th March 15 minutes
  Mathematics: Differentiation, 3rd April 15 minutes
  ... and then at monthly intervals.

Chapter 1

Core 3 and Core 4 (Pure Mathematics)

The following topics are covered in this chapter:

- Algebra and series
- Functions including exponential and logarithmic
- Coordinate geometry
- Trigonometry

- Differentiation
- Integration
- Numerical methods
- Vectors

1.1 Algebra and series

After studying this section you should be able to:

- simplify rational expressions
- express rational expressions in partial fractions
- use the binomial expansion $(1 + x)^n$ for any rational n

LEARNING SUMMARY

Rational expressions

Core 4

A **rational expression** is of the form $\dfrac{f(x)}{g(x)}$, where $f(x)$ and $g(x)$ are polynomials in x.

To simplify a rational expression, factorise both $f(x)$ and $g(x)$ as far as possible, then cancel any factors that appear in both the numerator and denominator.

Examples

(a) $\dfrac{2x^2 + 6x}{2x^2 + 7x + 3} = \dfrac{2x(x + 3)}{(2x + 1)(x + 3)}$

$= \dfrac{2x}{2x + 1}$

> This cannot be cancelled any further.

> Factorise $x^2 - 4$ using difference between two squares.

(b) $\dfrac{3x^2 - 12}{2 - x} = \dfrac{3(x^2 - 4)}{2 - x}$

> $x - 2 = -(2 - x)$.

$= \dfrac{3(x - 2)(x + 2)}{(2 - x)}$

$= -3(x + 2)$

When adding and subtracting algebra fractions, always use the lowest common denominator, for **example**,

> First factorise the denominators.

> The lowest common multiple of the denominators is $(x + 3)(x - 4)(x - 3)$.

$\dfrac{x - 2}{x^2 - x - 12} + \dfrac{4}{x^2 - 9} = \dfrac{x - 2}{(x + 3)(x - 4)} + \dfrac{4}{(x + 3)(x - 3)}$

$= \dfrac{(x - 3)(x - 2) + 4(x - 4)}{(x + 3)(x - 4)(x - 3)}$

$= \dfrac{x^2 - 5x + 6 + 4x - 16}{(x + 3)(x - 4)(x - 3)}$

$= \dfrac{x^2 - x - 10}{(x + 3)(x - 4)(x - 3)}$

Key points from AS

Remainder theorem page 18
Algebraic division page 17

Partial fractions

Core 4

It is sometimes possible to decompose a rational function into **partial fractions**. This format is often useful when differentiating, integrating or expanding functions as a series. Look out for the following types of partial fractions.

Denominator contains linear factors only

$$\frac{x+7}{(x-2)(x+1)} \equiv \frac{A}{x-2} + \frac{B}{x+1} = \frac{A(x+1)+B(x-2)}{(x-2)(x+1)}$$

$$\Rightarrow x+7 \equiv A(x+1)+B(x-2)$$

Let $x=-1$, then $6=-3B \Rightarrow B=-2$
Let $x=2$, then $9=3A \Rightarrow A=3$

The symbol $\equiv$ indicates that this is an identity. It is true for all values of x.

These substitutions are chosen so that each factor in turn becomes zero.

Examination questions could have up to three linear factors in the denominator.

$$\therefore \frac{x+7}{(x-2)(x+1)} \equiv \frac{3}{x-2} - \frac{2}{x+1}$$

Denominator contains repeated linear factors

Do not use $(x+3)(x+1)(x+1)^2$ as the common denominator as it is not the lowest common multiple.

$$\frac{6x^2+15x+7}{(x+3)(x+1)^2} \equiv \frac{A}{x+3} + \frac{B}{x+1} + \frac{C}{(x+1)^2}$$

$$\equiv \frac{A(x+1)^2+B(x+3)(x+1)+C(x+3)}{(x+3)(x+1)^2}$$

$$\Rightarrow 6x^2+15x+7 \equiv A(x+1)^2+B(x+3)(x+1)+C(x+3)$$

Use a mixture of substitution and equating coefficients to find A, B and C.

Let $x=-1$, then $-2=2C \Rightarrow C=-1$
Let $x=-3$, then $16=4A \Rightarrow A=4$
Equate x^2 terms: $6=A+B$ and since $A=4$, $B=2$.

$$\therefore \frac{6x^2+15x+7}{(x+3)(x+1)^2} \equiv \frac{4}{x+3} + \frac{2}{x+1} - \frac{1}{(x+1)^2}$$

The degree of a polynomial is its highest power of x, for example x^3+2x^2-3x+4 has degree 3.

Only **proper fractions** (when the numerator is of lower degree than the denominator) can be written in partial fraction form. For **improper fractions**, divide the denominator into the numerator first. This can be done by long division or by using identities.

Binomial expansion for rational n

Core 4

> When n is rational, $(1+x)^n$ can be written as a series, using the binomial expansion
>
> $$(1+x)^n = 1 + nx + \frac{n(n-1)}{2!}x^2 + \frac{n(n-1)(n-2)}{3!}x^3 + \dots$$

KEY POINT

Key points from AS

- **Binomial expansion when n is a positive integer**
  Revise AS page 44

If n is a positive integer, the series terminates at the term in x^n.

For all other values of n, the series is infinite and converges provided that $|x| < 1$.

Example

(a) Expand $\sqrt{1-4x}$ as an ascending series in x, as far as the term in x^3, giving the set of values of x for which the series is valid.

(b) By substituting $x = 0.01$ into the series expansion of $\sqrt{1-4x}$, find $\sqrt{96}$ correct to 4 decimal places.

> $1 - 4x = 1 + (-4x)$, so use $(1 + x)^n$ with $n = \frac{1}{2}$ and substituting $(-4x)$ for x.

(a) $\sqrt{1-4x} = (1-4x)^{\frac{1}{2}}$

$$= 1 + \tfrac{1}{2}(-4x) + \frac{(\frac{1}{2})(-\frac{1}{2})}{2!}(-4x)^2 + \frac{(\frac{1}{2})(-\frac{1}{2})(-\frac{3}{2})}{3!}(-4x)^3 + \dots$$

$$= 1 - 2x - 2x^2 - 4x^3 + \dots$$

The series is valid provided that $|4x| < 1$, i.e. $|x| < \frac{1}{4}$.

> You should also state the set of values of x for which the expansion is valid, even when it is not specifically requested in the question.

(b) Substituting $x = 0.01$ into $\sqrt{1-4x}$ gives $\sqrt{0.96} = \sqrt{\dfrac{96}{100}} = \dfrac{\sqrt{96}}{10}$.

So $\sqrt{96} = 10(1 - 2(0.01) - 2(0.01)^2 - 4(0.01)^3 + \dots)$

$$= 10(1 - 0.02 - 0.0002 - 0.000004 +)$$

$$= 9.7980 \ (4 \text{ d.p.})$$

The function to be expanded must be in the form $(1 + \dots)^n$ or $(1 - \dots)^n$.
You may have to do some careful algebraic manipulation to get it in this form.

Example

Give the first four terms in the series expansion of $\dfrac{1}{(2+x)}$.

$$\frac{1}{(2+x)} = \frac{1}{2\left(1 + \dfrac{x}{2}\right)}$$

$$= \frac{1}{2}\left(1 + \frac{x}{2}\right)^{-1}$$

> Alternatively,
> $(2+x)^{-1} = 2^{-1}\left(1 + \dfrac{x}{2}\right)^{-1}$.

$$= \frac{1}{2}\left(1 + (-1)\left(\frac{x}{2}\right) + \frac{(-1)(-2)}{2!}\left(\frac{x}{2}\right)^2 + \frac{(-1)(-2)(-3)}{3!}\left(\frac{x}{2}\right)^3 + \dots\right)$$

$$= \frac{1}{2}\left(1 - \frac{x}{2} + \frac{x^2}{4} - \frac{x^3}{8} + \dots\right)$$

$$= \frac{1}{2} - \frac{x}{4} + \frac{x^2}{8} - \frac{x^3}{16} + \dots$$

Restriction on x: $\left|\dfrac{x}{2}\right| < 1$, i.e. $|x| < 2$.

Progress check

Core 4

1 Simplify:

(a) $\dfrac{2x^2 + 3x}{2x^2 + x - 3}$ (b) $\dfrac{9x - x^3}{3x^2 + x^3}$

Core 4

2 Express as a single fraction in its simplest form:

(a) $\dfrac{3(x-4)}{(x+2)(x-1)} + \dfrac{2(x+1)}{x-1}$

(b) $\dfrac{x^2 - 3x + 2}{x^2 + 3x - 4} - \dfrac{1}{(x+4)^2}$

Core 4

3 $\dfrac{4x^2 + x + 1}{(x+2)(x-3)(x+1)} \equiv \dfrac{A}{x+2} + \dfrac{B}{x-3} + \dfrac{C}{x+1}$

Find the values of A, B and C.

Core 4

4 Express in partial fractions:

$\dfrac{3x^2 + 4x + 5}{(x-1)^2(x+2)}$

Core 4

5 Write as a series in ascending powers of x as far as the term in x^3, stating the values of x for which the expansion is valid:

(a) $\dfrac{1}{1+x}$ (b) $(1 - 3x)^{-2}$ (c) $\sqrt{4 + x}$

(c) $2 + \frac{1}{4}x - \frac{1}{64}x^2 + \frac{1}{512}x^3 + \cdots$ for $|x| < 4$

(b) $1 + 6x + 27x^2 + 108x^3 + \cdots$ for $|x| < \frac{1}{3}$

5 (a) $1 - x + x^2 - x^3 + \cdots$ for $|x| < 1$

4 $\dfrac{2}{x-1} + \dfrac{4}{(x-1)^2} + \dfrac{1}{x+2}$

3 $A = 3$, $B = 2$, $C = -1$

2 (a) $\dfrac{2x^2 + 9x - 8}{(x+2)(x-1)}$ (b) $\dfrac{x^2 + 2x - 9}{(x+4)^2}$

1 (a) $\dfrac{x}{x-1}$ (b) $\dfrac{3-x}{x}$

1.2 Functions including exponential and logarithmic

After studying this section you should be able to:

- *use function notation and understand domain and range*
- *understand composite functions and find the inverse of a one-one function*
- *sketch the graph of the inverse of a function from the graph of the function*
- *understand the properties of the modulus function*
- *use transformations to sketch graphs of related functions*
- *understand the definitions of e^x and ln x*

LEARNING SUMMARY

Functions

Core 3

A **function** may be thought of as a rule which takes each member x of a set and assigns, or **maps**, it to some value y known as its **image**.

> x maps to y
> y is the image of x.

$$x \longrightarrow \boxed{\text{Function}} \longrightarrow y$$

> $f(x)$ is read as 'f of x'.

A letter such as f, g or h is often used to stand for a function. The function which squares a number and adds on 5, for example, can be written as $f(x) = x^2 + 5$. The same notation may also be used to show how a function affects particular values. For this function, $f(4) = 4^2 + 5 = 21$, $f(-10) = (-10)^2 + 5 = 105$ and so on.

An alternative notation for the same function is $f: x \mapsto x^2 + 5$.

> **KEY POINT**
>
> The set of values on which the function acts is called the **domain** and the corresponding set of image values is called the **range**.

The function $f(x) = x + 2$ with domain {3, 6, 10} is shown below.

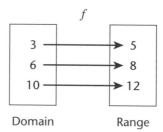

Domain   Range

> The symbol $\mathbb{R}$ is often used to stand for the complete set of numbers on the number line. This is the set of Real numbers.

The domain of a function may be an infinite set such as $\mathbb{R}$ but, in some cases, particular values must be omitted for the function to be valid.

For example, the function $f(x) = \dfrac{x + 2}{x - 3}$ cannot have 3 in its domain since division by zero is undefined.

Under a function, every member of the domain must have only one image. But, it is possible for different values in the domain to have the same image.

> There are different types of function.

> **KEY POINT**
>
> If for each element y in the range, there is *a unique* value of x such that $f(x) = y$ then f is a **one–one** function. If for any element y in the range, there is more than one value of x satisfying $f(x) = y$ then f is **many–one**.

Examples

The function $f(x) = 2x$ with domain $\mathbb{R}$ is one–one.

The function $f(x) = x^2$ with domain $\mathbb{R}$ is many–one.

The function $f(x) = x^2$ with domain $x > 0$ is one–one.

The function $f(x) = \sin x$ with domain $0° \leqslant x \leqslant 360°$ is many–one.

The function $f(x) = \sin x$ with domain $-90° \leqslant x \leqslant 90°$ is one–one.

> These examples show the importance of including the domain of the function in its definition.

Composition of functions

Core 3

The diagram shows how two functions f and g may be combined.

$$x \xrightarrow{\;f\;} f(x) \xrightarrow{\;g\;} g(f(x))$$

In this case, f is applied first to some value x giving $f(x)$. Then g is applied to the value $f(x)$ to give $g(f(x))$. This is usually written as $gf(x)$ and gf can be thought of as a new **composite function** defined from the functions f and g.

For example, if $f(x) = 3x$, $x \in \mathbb{R}$ and $g(x) = x + 2$, $x \in \mathbb{R}$ then
$$gf(x) = g(3x) = 3x + 2.$$

The order in which the functions are applied is important. The composite function fg is found by applying g first and then f.

In this case, $fg(x) = f(x + 2) = 3x + 6.$

Generally speaking, when two functions f and g are defined, the composite functions fg and gf will not be the same.

The inverse of a function

Core 3

The **inverse of a function** f is a function, usually written as f^{-1}, that *undoes* the effect of f. So the inverse of a function which adds 2 to every value, for example, will be a function that subtracts 2 from every value.

This can be written as $f(x) = x + 2$, $x \in \mathbb{R}$ and $f^{-1}(x) = x - 2$, $x \in \mathbb{R}$.

The domain of f^{-1} is given by the range of f.

Notice that $f^{-1}f(x) = f^{-1}(x + 2) = x$ and that $ff^{-1}(x) = f(x - 2) = x$.

A function can be either one–one or many–one, but only functions that are one–one can have an inverse. The reason is, that reversing a many–one function would give a mapping that is one–many, and this cannot be a function.

The diagram shows a many–one function. It does not have an inverse.

> Every value in the domain of a function can have only one image. This is an important property of functions.

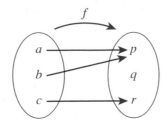

If you reverse this diagram then p would have more than one image.

Some important examples of restricting the domain of a function so that it will have an inverse are found in trigonometry. See page 28.

You can turn a many–one function into a one–one function by restricting its domain.

For example, the function $f(x) = x^2$, $x \in \mathbb{R}$ is many–one and so it does not have an inverse. However, the function $f(x) = x^2$, $x > 0$ is one–one and $f^{-1}(x) = \sqrt{x}$, $x > 0$.

Finding the inverse of a one–one function

Core 3

One way to find the inverse of a one–one function f is to write $y = f(x)$ and then rearrange this to make x the subject so that $x = f^{-1}(y)$. The inverse function is then usually defined in terms of x to give $f^{-1}(x)$. The domain of f^{-1} is the range of f.

You can turn $f^{-1}(y)$ into $f^{-1}(x)$ by replacing every y with an x.

Example   Find the inverse of the function $f(x) = \dfrac{x+2}{x-3}$, $x \neq 3$.

Start by writing $y = f(x)$.

Define $y = \dfrac{x+2}{x-3}$

Rearrange to find x.

then $y(x-3) = x+2$

Multiply out the brackets.

$xy - 3y = x + 2$

Collect the x terms on one side of the equation.

$xy - x = 3y + 2$

$x(y-1) = 3y + 2$

$x = \dfrac{3y+2}{y-1}.$

$x = f^{-1}(y)$ so this defines the inverse function in terms of y.

It is usual to express the inverse function in terms of x.

So $f^{-1}(x) = \dfrac{3x+2}{x-1}$, $x \neq 1$.

The denominator cannot be allowed to be zero so 1 is not in the domain of f^{-1}.

Here are the graphs of $y = \dfrac{x+2}{x-3}$ and its inverse $y = \dfrac{3x+2}{x-1}$

The graph of the inverse function is the reflection of the original graph in the line $y = x$. This is true for any inverse function but the same scale must be used on both axes or the effect is distorted.

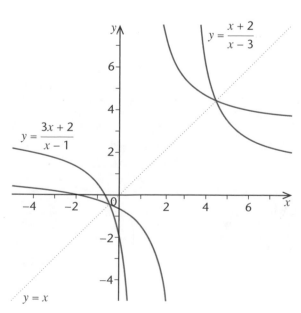

$y = \dfrac{x+2}{x-3}$

$y = \dfrac{3x+2}{x-1}$

$y = x$

The modulus function

Core 3

The notation $|x|$ is used to stand for the modulus of x. This is defined as

$$|x| = \begin{cases} x & \text{when } x \geqslant 0 \quad \text{(when } x \text{ is positive, } |x| \text{ is just the same as } x\text{).} \\ -x & \text{when } x < 0 \quad \text{(when } x \text{ is negative, } |x| \text{ is the same as } -x\text{).} \end{cases}$$

The expression $|x-a|$ can be interpreted as the distance between the numbers x and a on the number line. In this way, the statement $|x-a| < b$ means that the distance between x and a is less than b.

It follows that $a - b < x < a + b$.

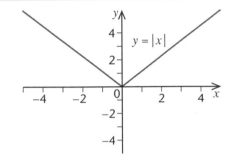

The graph of $y = |x|$

It follows that the graph of $y = |x|$ is the same as the graph of $y = x$ for positive values of x. But, when x is negative, the corresponding part of the graph of $y = x$ must be reflected in the x-axis to give the graph of $y = |x|$.

$|x|$ is never negative so the graph of $y = |x|$ doesn't go below the x-axis anywhere.

The graph of $y = |f(x)|$

The graph of $y = |f(x)|$ is the same as the graph of $y = f(x)$ for positive values of $f(x)$. But, when $f(x)$ is negative, the corresponding part of the graph of $y = f(x)$ must be reflected in the x-axis to give the graph of $y = |f(x)|$.

The graph of $y = 1/x$ goes below the x-axis for negative values of x.

The diagram shows the graph of $y = \left| \dfrac{1}{x} \right|$.

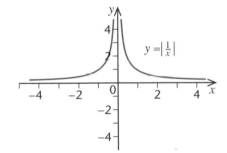

Transforming graphs

Core 3

The graph of some new function can often be obtained from the graph of a known function by applying a transformation. A summary of the standard transformations is given in the table.

You may need to apply a combination of transformations in some cases.

Known function	New function	Transformation
$y = f(x)$	$y = f(x) + a$	Translation through a units parallel to y-axis.
	$y = f(x - a)$	Translation through a units parallel to x-axis.
	$y = af(x)$	One-way stretch with scale factor a parallel to the y-axis.
	$y = f(ax)$	One-way stretch with scale factor $\frac{1}{a}$ parallel to the x-axis.

Examples

$y = af(x)$ does not move points on the *x*-axis.

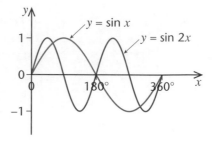

Sometimes you may have to perform more than one transformation.

Example

Describe a sequence of two transformations that maps the graph of $y = \cot x$ onto the graph of $y = 3 + \cot 2x$.

Stretch by scale factor $\frac{1}{2}$ parallel to the *x*-axis and translate using the vector $\begin{bmatrix} 0 \\ 3 \end{bmatrix}$.

The exponential function e^x

Core 3

An **exponential function** is one where the variable is a power or exponent. For example, any function of the form $f(x) = a^x$, where a is a constant, is an exponential function.

The diagram shows some graphs of exponential functions for different values of a.

All of the graphs pass through the point (0, 1) and each one has a different gradient at this point. The value of the gradient depends on a.

Exponential functions are often used to represent, or model, patterns of:
- growth when $a > 1$
- decay when $a < 1$.

Key points from AS

Exponential functions
Revise AS page 39

All scientific calculators will display values of e^x and 10^x. To display the value of e, for example, find e^1.

There is a particular value of a for which the curve has gradient 1 at (0, 1). The letter e is used to stand for this value and the function $f(x) = e^x$ is often called **the exponential function** because of its special importance to the subject.
e = 2.718281828 ... It is an irrational number and so it cannot be written exactly.

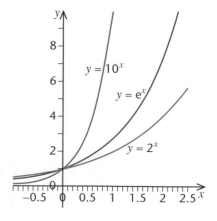

You may be asked to sketch other curves relating to e^x, for example

Notice $e^x > 0$ for all values of x.

Reflection in y-axis.

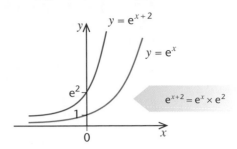

$e^{x+2} = e^x \times e^2$

Translation 2 units left
or stretch, factor e^2, parallel to y-axis.

The natural logarithmic function, ln x

Core 3

If $y = e^x$ then $x = \log_e y$. This is usually written as $x = \ln y$.

Key points from AS

Logarithms Revise AS page 39

Logarithms to the base e are called *natural logarithms*.

Your calculator has a key for natural logs, probably labelled ln.

$e^0 = 1 \Rightarrow \ln 1 = 0$
$e^1 = e \Rightarrow \ln e = 1$

In fact, $\ln e^{f(x)} = f(x)$, for example $\ln e^{x^2} = x^2$, $\ln e^{\sin x} = \sin x$.

Since the exponential and natural logarithm functions are inverses of each other, their graphs are symmetrical about the line $y = x$.

This is true for logs of any base.

The diagram shows that the domain of the natural logarithm function is given by $x > 0$.

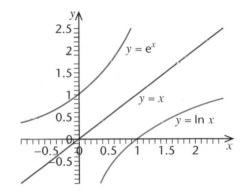

You may be asked to sketch other curves relating to $y = \ln x$, for example

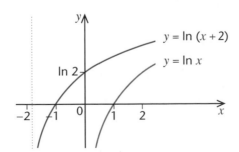

Translation 2 units to the left.

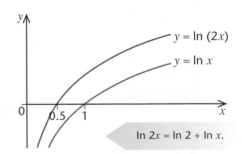

$\ln 2x = \ln 2 + \ln x$.

Stretch parallel to the x-axis factor $\frac{1}{2}$
or translation ln 2 units up.

Progress check

Core 3   **1** $f(x) = x^2 + 5$ and $g(x) = 2x + 3$.

     (a) Write $fg(x)$ in terms of x.   (b) Find $fg(10)$.

     (c) Find the values of x for which $fg(x) = gf(x)$.

Core 3   **2**   Find the inverse of the function $f(x) = \dfrac{x+5}{x-2}$, $x \in \mathbb{R}$, $x \ne 2$ and state the domain of the inverse function.

Core 3   **3**   (a) Sketch $y = (x-2)(x+2)$.

     (b) Sketch $y = |x^2 - 4|$.

Core 3   **4**   Solve $|x-2| < 5$.

Core 3   **5**   The diagram shows $y = f(x)$.
On separate diagrams, sketch
     (a) $y = f(x) + 1$
     (b) $y = f(x+1)$
     (c) $y = f(2x)$
     (d) $y = 2f(x)$
     (e) $y = f(\tfrac{1}{2}x)$

Core 3   **6**   On the same diagram, sketch $y = x^2$, $y = (x-2)^2$, $y = -x^2$.

Core 3   **7**   On the same diagram, sketch $y = e^x$, $y = e^{2x}$, $y = -e^x$.

7 $y = e^x$, $y = e^{2x}$, $y = -e^x$

6 $y = x^2$, $y = (x-2)^2$, $y = -x^2$

5   (a) (b) (c) (d) (e)

4 $-3 < x < 7$

3   (a) (b)

2 $f(x) = \dfrac{2x+5}{x-1}$, $x \in \mathbb{R}$, $x \ne 1$.

1   (a) $fg(x) = 4x^2 + 12x + 14$ (b) $fg(10) = 534$ (c) $x = -0.08452, -5.915$.

1.3 Coordinate geometry

After studying this section you should be able to:

- sketch curves in Cartesian and parametric form
- convert the equation of a curve between Cartesian and parametric form

LEARNING SUMMARY

Cartesian and parametric form

Core 4

Cartesian form

A curve in Cartesian form is defined in terms of x and y only, **for example**
$$y = x^2 + 3x \text{ in which } y \text{ is given } \textbf{explicitly} \text{ in terms of } x.$$
$$x^2 + 2xy + y^2 = 9 \text{ in which the equation is given } \textbf{implicitly}.$$

To sketch a function given in Cartesian form:

Key points from AS

- **Curve sketching**
  Revise AS page 30

If all the powers of x are even, the graph is symmetrical in the y-axis.

Look for values that make the denominator zero.

1   Find where the curve **crosses the axes** by putting $x = 0$ and $y = 0$.
2   Check for **symmetry**.
3   Check for **stationary points**.
4   Check what happens for **large values of x and y**.
5   Check for **discontinuities**.

Find when $\dfrac{dy}{dx} = 0$ and then investigate.

Parametric form

The parameter θ is often used when trigonometric functions are involved.

The curve is plotted by finding the values of x and y for various values of the parameter.

A curve is defined in **parametric form** by expressing x and y in terms of a third variable, for example $x = f(t)$, $y = g(t)$. In this case, t is the parameter.

Example
Plot the curve $x = 2t^2$, $y = 4t$ for $-3 \leqslant t \leqslant 3$ and find the Cartesian equation of the curve.

The curve is a parabola.

t	-3	-2	-1	0	1	2	3
x	18	8	2	0	2	8	18
y	-12	-8	-4	0	4	8	12

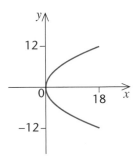

Eliminate t to obtain the Cartesian equation.

$$y = 4t \Longrightarrow t = \frac{y}{4}$$

Substituting into $x = 2t^2$ gives $x = 2\left(\dfrac{y}{4}\right)^2 = \dfrac{y^2}{8}$ so $y^2 = 8x$.

Key points from AS

- **Trigonometric identities**
  Revise AS page 49

When the parameter is θ, it may be necessary to use a trigonometric identity to find the Cartesian equation.

Key points from AS

• **Circle**
  Revise AS page 26

For example, when $x = a \cos \theta$ and $y = a \sin \theta$, use $\cos^2 \theta + \sin^2 \theta = 1$. This gives $\dfrac{x^2}{a^2} + \dfrac{y^2}{a^2} = 1$ or $x^2 + y^2 = a^2$.

This is a circle, centre $(0, 0)$, radius a.

Also, when $x = a \cos \theta$, $y = b \sin \theta$, $\dfrac{x^2}{a^2} + \dfrac{y^2}{b^2} = 1$.

This curve is an ellipse.

Progress check

Core 4

1 A curve has parametric equations $x = t - 1$, $y = t^2 + t^3$. Find the co-ordinates of the points where the curve crosses the axes.

Core 4

2 A curve is defined parametrically by $x = 2t$, $y = \dfrac{2}{t}$.

(a) Give the Cartesian equation of the curve.

(b) Sketch the curve.

Core 4

3 Describe the curve defined parametrically by $x = 5 \cos \theta$, $y = 5 \sin \theta$ and give its Cartesian equation.

3 Circle, centre $(0, 0)$, radius 5; $x^2 + y^2 = 25$.

(b)
2 (a) $xy = 4$ (hyperbola)
1 $(-1, 0)$, $(-2, 0)$, $(0, 2)$

1.4 Trigonometry

Secant, cosecant and cotangent

Core 3

These trigonometric functions, commonly known as **sec**, **cosec** and **cot**, are defined from the more familiar sin, cos and tan functions as follows:

$$\sec x = \frac{1}{\cos x} \qquad \operatorname{cosec} x = \frac{1}{\sin x} \qquad \cot x = \frac{1}{\tan x}$$

Key points from AS

- **Trigonometry** pages 46–50

Each function is periodic and has either line or rotational symmetry. The domain of each one must be restricted to avoid division by zero.
You need to be able to work in degrees or radians.
Remember $180° = \pi$ radians.

$y = \sec x$

- The period of sec is 360° (2π radians) to match the period of cos.
- Notice that sec x is undefined whenever cos $x = 0$.
- The graph is symmetrical about every vertical line passing through a vertex.
- It has rotational symmetry of order 2 about the points in the x-axis corresponding to $90° \pm 180°n$ ($\frac{\pi}{2} \pm \pi n$).

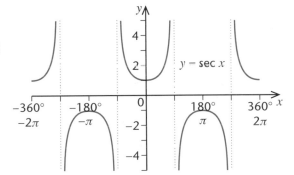

$y = \operatorname{cosec} x$

- The period of cosec is 360° (2π radians) to match the period of sin.
- Notice that cosec x is undefined whenever sin $x = 0$.
- The graph is symmetrical about every vertical line passing through a vertex.
- It has rotational symmetry of order 2 about the points on the x-axis corresponding to 0°, $\pm 180°$, $\pm 360°$, ..., $(0, \pm \pi, \pm 2\pi, ...)$.

$y = \cot x$

$$\frac{1}{\tan x} = \frac{\cos x}{\sin x}.$$

- The period of cot is 180° to match the period of tan.
- Notice that cot x is undefined whenever sin $x = 0$.
- The graph has rotational symmetry of order 2 about the points on the x-axis corresponding to 0°, $\pm 90°$, $\pm 180°$, ..., $(0, \pm \frac{\pi}{2}, \pm \pi, ...)$.

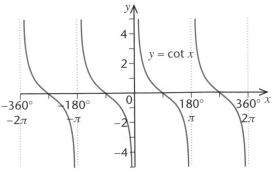

Inverse trigonometric functions

Core 3

The sine, cosine and tangent functions are all many–one and so do not have inverses on their full domains. However, it is possible to restrict their domains so that each one has an inverse. The graphs of these inverse functions are given below.

See page 19 to find out about inverse functions.

You need to know these graphs and to understand the need for restricting the domains.

$f(x) = \sin x, \; -\frac{\pi}{2} \leqslant x \leqslant \frac{\pi}{2}$

$(-90° \leqslant x \leqslant 90°)$

$\Rightarrow f^{-1}(x) = \sin^{-1} x, \; -1 \leqslant x \leqslant 1.$

$f(x) = \cos x, \; 0 \leqslant x \leqslant \pi$

$(0° \leqslant x \leqslant 180°)$

$\Rightarrow f^{-1}(x) = \cos^{-1} x, \; -1 \leqslant x \leqslant 1.$

$f(x) = \tan x, \; -\frac{\pi}{2} < x < \frac{\pi}{2}$

$(-90° < x < 90°)$

$\Rightarrow f^{-1}(x) = \tan^{-1} x, \; x \in \mathbb{R}.$

This is the notation used for inverse trig' functions on your calculator.

But some texts use:

arcsin for $\sin^{-1}$

arccos for $\cos^{-1}$

arctan for $\tan^{-1}$.

$\sin^{-1} x$ means:

'the angle whose sine is x'

e.g. $\sin^{-1} 0.5 = 30°$.

In radians $\sin^{-1} 0.5 = \frac{\pi}{6}$.

$\cos^{-1} x$ means:

'the angle whose cosine is x'

e.g. $\cos^{-1} 0.5 = 60°$.

In radians $\cos^{-1} 0.5 = \frac{\pi}{3}$.

$\tan^{-1} x$ means:

'the angle whose tangent is x'

e.g. $\tan^{-1} 1 = 45°$.

In radians $\tan^{-1} 1 = \frac{\pi}{4}$.

When you use the functions $\sin^{-1}$, $\cos^{-1}$ and $\tan^{-1}$ on your calculator, the value given is called the **principal value** (PV).

Trigonometric identities

You should *learn* the identities listed below. It is so much easier to simplify expressions, establish new identities and solve equations if you have a firm grasp of these basic results.

Core 3

- The **Pythagorean identities** are:

$$\cos^2 \theta + \sin^2 \theta \equiv 1$$

$$1 + \tan^2 \theta \equiv \sec^2 \theta$$

$$\cot^2 \theta + 1 \equiv \operatorname{cosec}^2 \theta$$

The second result can be obtained from the first by dividing throughout by $\cos^2\theta$.
The third result can be obtained in a similar way, by dividing throughout by $\sin^2\theta$.

Core 4

- The **compound angle identities** for $(A + B)$ are:

$$\sin(A + B) \equiv \sin A \cos B + \cos A \sin B$$

$$\cos(A + B) \equiv \cos A \cos B - \sin A \sin B$$

$$\tan(A + B) \equiv \frac{\tan A + \tan B}{1 - \tan A \tan B}.$$

The corresponding results for $(A - B)$ are:

$$\sin(A - B) \equiv \sin A \cos B - \cos A \sin B$$

$$\cos(A - B) \equiv \cos A \cos B + \sin A \sin B$$

$$\tan(A - B) \equiv \frac{\tan A - \tan B}{1 + \tan A \tan B}.$$

These results can be worked out from the ones for $(A + B)$. Just replace every '+' with a '−' and vice versa. This makes it easier to learn the results.

You can obtain these results from the compound angle formulae by setting $A = B$.

- The **double angle identities** are:

$$\sin 2A \equiv 2\sin A \cos A$$

$$\cos 2A \equiv \cos^2 A - \sin^2 A$$

$$\equiv 2\cos^2 A - 1$$

$$\equiv 1 - 2\sin^2 A$$

These variations are very useful – it is worth knowing them.

$$\tan 2A \equiv \frac{2\tan A}{1 - \tan^2 A}.$$

- The **half angle identities** are:

These results are useful in integration.

$$\sin^2 \tfrac{1}{2}A \equiv \tfrac{1}{2}(1 - \cos A)$$

$$\cos^2 \tfrac{1}{2}A \equiv \tfrac{1}{2}(1 + \cos A).$$

Proving identities

The basic technique for proving a given identity is:

- Start with the expression on one side of the identity.
- Use known identities to replace some part of it with an equivalent form.
- Simplify the result and compare with the expression on the other side.

This is a skill. The more you do, the better you will become.

With practice you get to know which parts to replace so that the required result is established.

Core 4

Example

Prove that $\dfrac{\cos 2A}{\cos A - \sin A} \equiv \cos A + \sin A$.

Starting with the LHS:

$$\frac{\cos 2A}{\cos A - \sin A} \equiv \frac{\cos^2 A - \sin^2 A}{\cos A - \sin A}$$

Using one of the double angle results.

$$\equiv \frac{(\cos A + \sin A)(\cos A - \sin A)}{\cos A - \sin A}$$

Using the difference of two squares.

$$\equiv \cos A + \sin A \equiv \text{RHS.}$$

Cancelling the common factor.

It is often a good idea to replace sec, cosec and cot with sin, cos and tan at the start.

Core 3

Example

Prove that $\sec^2 A \equiv \dfrac{\operatorname{cosec} A}{\operatorname{cosec} A - \sin A}$.

It's often easier to start with the more complicated looking side and try to simplify it.

$$\text{RHS} \equiv \frac{\dfrac{1}{\sin A}}{\dfrac{1}{\sin A} - \sin A} \equiv \frac{1}{1 - \sin^2 A}$$

Replacing cosec A with $\dfrac{1}{\sin A}$ and multiplying the numerator and denominator by $\sin A$.

$$\equiv \frac{1}{\cos^2 A} \equiv \sec^2 A \equiv \text{LHS.}$$

Using one of the Pythagorean identities and the definition of sec A.

Trigonometric equations

Core 3 & 4

Identities may be used to re-write an equation in a form that is easier to solve.

Core 4

Example Solve the equation $\cos 2x + \sin x + 1 = 0$ for $-\pi \leqslant x \leqslant \pi$.

Replacing $\cos 2x$ with $1 - 2\sin^2 x$ gives $1 - 2\sin^2 x + \sin x + 1 = 0$

$$\Rightarrow 2\sin^2 x - \sin x - 2 = 0.$$

> You need to recognise that this is a quadratic in $\sin x$.

Using the quadratic formula gives $\sin x = \dfrac{1 \pm \sqrt{1 + 16}}{4} = \dfrac{1 \pm \sqrt{17}}{4}$.

$$\Rightarrow \sin x = 1.28 \ldots \text{ (no solutions) or } \sin x = -0.78077 \ldots.$$

Using $\sin^{-1}$ gives $\qquad x = -0.8959$ radians to 4 d.p.

> Set calculator to radian mode.

> $-1 \leqslant \sin x \leqslant 1$ for all values of x.

Another solution in the interval $-\pi \leqslant x \leqslant \pi$ is $-\pi + 0.8959$ giving $x = -2.2457$ radians to 4 d.p.

The diagram shows there are no other solutions in this interval.

> **KEY POINT**
>
> An important result is that the expression $a\cos\theta + b\sin\theta$ can be written in any of the equivalent forms $r\cos(\theta \pm \alpha)$ or $r\sin(\theta \pm \alpha)$ for a suitable choice of r and α. The convention is to take $r > 0$.

One application of the result is to solve equations of the form $a\cos\theta + b\sin\theta = c$.

Core 4

Example Express $3\cos\theta - 4\sin\theta$ in the form $r\cos(\theta + \alpha)$, $r > 0$. Use the result to solve the equation $3\cos\theta - 4\sin\theta = 2$ for $-\pi \leqslant \theta \leqslant \pi$.

> You need to expand $r\cos(\theta + \alpha)$ and match corresponding parts of the identity.

$$r\cos(\theta + \alpha) \equiv 3\cos\theta - 4\sin\theta$$

$$\Rightarrow r\cos\theta\cos\alpha - r\sin\theta\sin\alpha \equiv 3\cos\theta - 4\sin\theta$$

$$\Rightarrow r\cos\alpha = 3 \text{ and } r\sin\alpha = 4$$

> $\dfrac{r\sin\alpha}{r\cos\alpha} = \dfrac{4}{3}$
>
> $\Rightarrow \tan\alpha = \dfrac{4}{3}$

$$\Rightarrow r = \sqrt{3^2 + 4^2} = 5 \text{ and } \alpha = \tan^{-1}\left(\tfrac{4}{3}\right) = 0.9273^c \text{ to 4 d.p.}$$

$$\Rightarrow 3\cos\theta - 4\sin\theta \equiv 5\cos(\theta + 0.9273^c).$$

Using this result in the equation $3\cos\theta - 4\sin\theta = 2$ gives

$$5\cos(\theta + 0.9273^c) = 2$$

$$\Rightarrow \cos(\theta + 0.9273^c) = 0.4.$$

This gives $\theta + 0.9273^c = \pm 1.1593^c \Rightarrow \theta = 0.232^c$ or $\theta = -2.087^c$ to 3 d.p.

Progress check

Core 4

1 Prove the identity $\sin 2\theta \equiv \dfrac{2\tan\theta}{1 + \tan^2\theta}$.

Core 4

2 Solve the equation $\cos 2x - \sin x = 0$ for $0° \leqslant x \leqslant 360°$.

Core 4

3 (a) Write $5\sin x + 12\cos x$ in the form $r\sin(x + \alpha)$ where $0 < \alpha < \dfrac{\pi}{2}$.

   (b) Solve the equation $5\sin x + 12\cos x = 7$ for $-\pi \leqslant x \leqslant \pi$.

3 (a) $13\sin(x + 1.176^c)$ (b) $x = -0.607^c$ or $x = 1.397^c$ to 3 d.p.

2 $x = 30°, 150°$ or $270°$

1 RHS $= \dfrac{2\tan\theta}{\sec^2\theta} \equiv 2\tan\theta \times \cos^2\theta \equiv 2\sin\theta\cos\theta \equiv \sin 2\theta \equiv$ LHS.

1.5 Differentiation

After studying this section you should be able to:

- use the chain rule to differentiate composite functions
- differentiate functions based on e^x and $\ln x$
- differentiate trigonometric functions
- differentiate using the product and quotient rules
- understand and use the relation $\dfrac{dy}{dx} = 1 \div \dfrac{dx}{dy}$
- find derivatives of functions defined parametrically and implicitly

LEARNING SUMMARY

The chain rule

Core 3

> **KEY POINT**
>
> The **chain rule** can be written as $\dfrac{dy}{dx} = \dfrac{dy}{du} \times \dfrac{du}{dx}$, where u is a function of x.

It shows how to differentiate a composite function by separating the process into simpler stages and combining the results.

Example Differentiate   (a) $y = (3x + 2)^{10}$

                    (b) $y = \sqrt{x^2 - 9}$.

Key points from AS

- **Differentiation**
  Revise AS pages 29–31

(a) Taking $u = 3x + 2$ gives $y = u^{10}$ and $\dfrac{dy}{du} = 10u^9$, so $\dfrac{dy}{du} = 10(3x + 2)^9$.

It also gives $\dfrac{du}{dx} = 3$.

So, from the chain rule: $\dfrac{dy}{dx} = 10(3x + 2)^9 \times 3 = 30(3x + 2)^9$.

(b) Taking $u = x^2 - 9$ gives $y = \sqrt{u} = u^{\frac{1}{2}}$, and $\dfrac{dy}{du} = \frac{1}{2} u^{-\frac{1}{2}}$, so $\dfrac{dy}{du} = \frac{1}{2}(x^2 - 9)^{-\frac{1}{2}}$.

It also gives $\dfrac{du}{dx} = 2x$.

So, from the chain rule: $\dfrac{dy}{dx} = \frac{1}{2}(x^2 - 9)^{-\frac{1}{2}} \times 2x = \dfrac{x}{\sqrt{x^2 - 9}}$.

The chain rule can also be used to establish results for **connected rates of change**.

The exponential function e^x

Core 3

The exponential function was defined on page 22. It has the special property that it is unchanged by differentiation. So, if $y = e^x$ then $\dfrac{dy}{dx} = e^x$.

Constant multiples of the exponential function behave in the same way, so if $y = ae^x$ then $\dfrac{dy}{dx} = ae^x$ where a is any constant.

Examples

Differentiate with respect to x:

(a) $y = 2e^x$ (b) $y = x^3 + e^x$ (c) $f(x) = \dfrac{1}{x} - 3e^x$

(a) $\dfrac{dy}{dx} = 2e^x$ (b) $\dfrac{dy}{dx} = 3x^2 + e^x$ (c) $f'(x) = -\dfrac{1}{x^2} - 3e^x$

You may need to use the chain rule.

Examples

Differentiate with respect to x:

(a) $y = e^{3x+2}$ (b) $y = e^{4x^2}$

(a) Let $u = 3x + 2$ so that $y = e^u$

$$\frac{du}{dx} = 3 \quad \text{and} \quad \frac{dy}{du} = e^u$$

Using the chain rule

$$\frac{dy}{dx} = \frac{dy}{du} \times \frac{du}{dx} = e^u \times 3 = 3\,e^{3x+2}$$

(b) Let $u = 4x^2$, so that $y = e^u$

$$\frac{du}{dx} = 8x \quad \text{and} \quad \frac{dy}{du} = e^u$$

$$\frac{dy}{dx} = \frac{dy}{du} \times \frac{du}{dx} = e^u \times 8x = 8xe^{4x^2}$$

In general, it can be shown using the chain rule

$f'(x)$ is the derivative, with respect to x, of $f(x)$.

> **KEY POINT**
>
> If $y = e^{f(x)}$ then $\dfrac{dy}{dx} = f'(x)e^{f(x)}$.

In particular, it is useful to remember that if $y = e^{ax+b}$, then $\dfrac{dy}{dx} = ae^{ax+b}$

Logarithmic functions ln x and ln $f(x)$

Core 3

> **KEY POINT**
>
> If $y = \ln x$ then $\dfrac{dy}{dx} = \dfrac{1}{x}$.

Key points from AS

- **Logarithms**
  Revise AS page 39

$\log a + \log b = \log(ab)$.
$\log x^k = k \log x$.

This result is easily extended to functions of the form $y = \ln(kx)$ and $y = \ln(x^k)$ using the laws of logarithms.

If $y = \ln(kx)$, then $y = \ln k + \ln x$

giving $\dfrac{dy}{dx} = \dfrac{1}{x} + 0 = \dfrac{1}{x}$.

Since $\ln k$ is a constant, its derivative is zero.

For example, if $y = \ln(5x)$, then $\dfrac{dy}{dx} = \dfrac{1}{x}$.

If $y = \ln(x^k)$, then $y = k \ln x$ giving $\dfrac{dy}{dx} = k \times \dfrac{1}{x} = \dfrac{k}{x}$.

For example, if $y = \ln\left(\dfrac{1}{x^2}\right) = \ln(x^{-2}) = -2 \ln x$, then $\dfrac{dy}{dx} = -2 \times \dfrac{1}{x} = -\dfrac{2}{x}$.

Always check whether a logarithmic function can be simplified before attempting to differentiate.

You may need to use the chain rule to differentiate $y = \ln f(x)$.

Example

Differentiate $y = \ln(x^2 + 3x)$.

This function cannot be simplified.

Let $u = x^2 + 3x$ so that $y = \ln u$

$$\dfrac{du}{dx} = 2x + 3 \quad \text{and} \quad \dfrac{dy}{du} = \dfrac{1}{u} = \dfrac{1}{x^2 + 3x}$$

Using the chain rule

$$\dfrac{dy}{dx} = \dfrac{dy}{du} \times \dfrac{du}{dx} = \dfrac{1}{x^2 + 3x} \times (2x + 3)$$

$$= \dfrac{2x + 3}{x^2 + 3x}$$

Notice that $2x + 3$ is the derivative of $x^2 + 3x$.

In general, it can be shown using the chain rule

This can be used directly without showing the chain rule working.

$$\dfrac{d}{dx} \ln(f(x)) = \dfrac{f'(x)}{f(x)}.$$

KEY POINT

Progress check

Find $\dfrac{dy}{dx}$ in each of the following questions.

Core 3   1 $y = (4 - 3x)^6$

Core 3   2 $y = \sqrt{2x + 1}$

Core 3   3 $y = e^{3x+1} + 3x$

Core 3   4 $y = \ln(7x) + \ln(x^4)$

Core 3   5 $y = \ln(x^2 - 3)$

5 $\dfrac{2x}{x^2 - 3}$

4 $\dfrac{5}{x}$

3 $3e^{3x+1} + 3$

2 $(2x+1)^{-\frac{1}{2}}$

1 $-18(4 - 3x)^5$

Trigonometric functions

Core 3

You should learn the derivatives of the three main **trigonometric functions**:

$$\frac{d}{dx}(\sin x) = \cos x \qquad \frac{d}{dx}\sin(ax+b) = a\cos(ax+b)$$

$$\frac{d}{dx}(\cos x) = -\sin x \qquad \frac{d}{dx}\cos(ax+b) = -a\sin(ax+b)$$

$$\frac{d}{dx}(\tan x) = \sec^2 x \qquad \frac{d}{dx}\tan(ax+b) = a\sec^2(ax+b)$$

> When differentiating, the angle must be in radians.

Example

Find the exact value of the gradient of $y = \cos 2x$ when $x = \dfrac{\pi}{6}$.

$$y = \cos 2x \implies \frac{dy}{dx} = -2\sin 2x$$

When $x = \dfrac{\pi}{6}$, $\dfrac{dy}{dx} = -2\sin\dfrac{\pi}{3} = -2 \times \dfrac{\sqrt{3}}{2} = -\sqrt{3}$

The **chain rule** can often be used to differentiate expressions involving trigonometric functions.

Example

Differentiate: (a) $y = 5\sin^4 x$ (b) $y = e^{\tan 4x}$ (c) $y = \ln(\sin x)$

> $\sin^4 x = (\sin x)^4$.

(a)   Let $u = \sin x$ so that $y = 5u^4$

Then $\dfrac{du}{dx} = \cos x$ and $\dfrac{dy}{du} = 20u^3 = 20(\sin x)^3 = 20\sin^3 x$

> With practice you should be able to do the working mentally.

By the chain rule:

$$\frac{dy}{dx} = \frac{dy}{du} \times \frac{du}{dx} = 20\sin^3 x \times (\cos x)$$

$$= 20\cos x \sin^3 x.$$

> Although not essential, it is usual to write an expression with the least complicated factors first.

(b)   Let $u = \tan 4x$ so that $y = e^u$

$$\frac{du}{dx} = 4\sec^2 4x \text{ and } \frac{dy}{du} = e^u = e^{\tan 4x}.$$

$$\therefore \frac{dy}{dx} = e^{\tan 4x} \times 4\sec^2 4x$$

$$= 4\sec^2 4x\, e^{\tan 4x}.$$

> In general, by the chain rule:
> $$\frac{d}{dx}(e^{f(x)}) = f'(x)e^{f(x)}.$$

(c) Use the result $y = \ln f(x) \implies \dfrac{dy}{dx} = \dfrac{f'(x)}{f(x)}$

> $f(x) = \sin x$, $f'(x) = \cos x$.

$$y = \ln(\sin x)$$

$$\frac{dy}{dx} = \frac{\cos x}{\sin x}$$

$$= \cot x.$$

The product rule

Core 3

Expressions written as the product of two functions, $f(x) \times g(x)$, can be differentiated using the **product rule**.

> **KEY POINT**
>
> If $y = uv$, where $u = f(x)$ and $v = g(x)$, then $\dfrac{dy}{dx} = u\dfrac{dv}{dx} + v\dfrac{du}{dx}$.

Example

Differentiate with respect to x: (a) $y = x^2 \sin 4x$ (b) $y = x^3 e^{-x}$

> With practice, the side working can be done mentally.

(a) Let $y = uv$, where $u = x^2$ and $v = \sin 4x$

$$\frac{dy}{dx} = u\frac{dv}{dx} + v\frac{du}{dx}$$

$$= x^2 \times 4\cos 4x + \sin 4x \times 2x$$

$$= 2x(2x\cos 4x + \sin 4x)$$

Side working

$$u = x^2 \implies \frac{du}{dx} = 2x$$

$$v = \sin 4x \implies \frac{dv}{dx} = 4\cos 4x$$

> Tidy the answer by taking out factors if possible.

(b) Let $y = uv$, where $u = x^3$ and $v = e^{-x}$

$$\frac{dy}{dx} = x^3 \times (-e^{-x}) + e^{-x} \times 3x^2$$

$$= x^2 e^{-x}(-x + 3)$$

Side working

$$u = x^3 \implies \frac{du}{dx} = 3x^2$$

$$v = e^{-x} \implies \frac{dv}{dx} = -e^{-x}$$

The quotient rule

Core 3

Expressions written in the form $\dfrac{f(x)}{g(x)}$ can be differentiated using the **quotient rule**.

> **KEY POINT**
>
> If $y = \dfrac{u}{v}$ where $u = f(x)$ and $v = g(x)$, then $\dfrac{dy}{dx} = \dfrac{v\dfrac{du}{dx} - u\dfrac{dv}{dx}}{v^2}$.

> Before using the quotient rule, check whether the expression can be simplified first, such as in
> $$y = \frac{4x^2 + 3x + 2}{x}.$$

Example

Find $\dfrac{dy}{dx}$ when $y = \dfrac{e^{2x}}{x^3}$.

Side working:

$$\frac{dy}{dx} = \frac{v\dfrac{du}{dx} - u\dfrac{dv}{dx}}{v^2}$$

$$= \frac{x^3(2e^{2x}) - e^{2x}(3x^2)}{(x^3)^2}$$

$$u = e^{2x} \implies \frac{du}{dx} = 2e^{2x}$$

$$v = x^3 \implies \frac{dv}{dx} = 3x^2$$

> Take out factors.

$$= \frac{x^2 e^{2x}(2x - 3)}{x^6}$$

> Simplify.

$$= \frac{e^{2x}(2x - 3)}{x^4}$$

Example

Find the coordinates of the stationary point on the curve $y = \dfrac{\ln x}{x}$.

Side working:

$$\frac{dy}{dx} = \frac{x \times \dfrac{1}{x} - \ln x \times 1}{x^2}$$

$$= \frac{1 - \ln x}{x^2}$$

$u = \ln x \Rightarrow \dfrac{du}{dx} = \dfrac{1}{x}$

$v = x \Rightarrow \dfrac{dv}{dx} = 1$

$\dfrac{dy}{dx} = 0$ when $\ln x = 1$, i.e. when $x = e$

In $e = 1$.

When $x = e$, $y = e^{-1}$, so the stationary point is at (e, e^{-1}).

Alternative formats of product and quotient rules when $u = f(x)$ and $v = g(x)$:

These formats are sometimes quoted. Learn and use the ones you find easiest to remember.

Product rule: If $y = f(x) \times g(x)$, then $\dfrac{dy}{dx} = f'(x)g(x) + f(x)g'(x)$

Quotient rule: If $y = \dfrac{f(x)}{g(x)}$, then $\dfrac{dy}{dx} = \dfrac{f'(x)g(x) - f(x)g'(x)}{[g(x)]^2}$

Using the result $\dfrac{dy}{dx} = 1 \div \dfrac{dx}{dy}$

Core 3

$$\frac{dy}{dx} = 1 \div \frac{dx}{dy} = \frac{1}{\dfrac{dx}{dy}}.$$

KEY POINT

For example, if $x = 4y + 2$, then $\dfrac{dx}{dy} = 4$ and $\dfrac{dy}{dx} = \dfrac{1}{\dfrac{dx}{dy}} = \dfrac{1}{4}$.

Parametric functions

Core 4

If x and y are each expressed in terms of a parameter t,

then $\dfrac{dy}{dx} = \dfrac{dy}{dt} \times \dfrac{dt}{dx}$.   Remember that $\dfrac{dt}{dx} = \dfrac{1}{dx/dt}$.

KEY POINT

Example

A curve is defined parametrically by $x = 3t^2$, $y = 4t$.

Find $\dfrac{dy}{dx}$ when $y = 8$

$$x = 3t^2 \Rightarrow \frac{dx}{dt} = 6t \Rightarrow \frac{dt}{dx} = \frac{1}{6t}$$

$$y = 4t \Rightarrow \frac{dy}{dt} = 4$$

$$\therefore \quad \frac{dy}{dx} = \frac{dy}{dt} \times \frac{dt}{dx} = 4 \times \frac{1}{6t} = \frac{2}{3t}$$

> Find the value of t when $y = 8$.

When $y = 8$, $4t = 8$, i.e. $t = 2$; when $t = 2$, $\dfrac{dy}{dx} = \dfrac{2}{3t} = \dfrac{2}{6} = \dfrac{1}{3}$.

Implicit functions

Core 4

> To find $\dfrac{dy}{dx}$ when an equation in x and y is given **implicitly**, differentiate each term with respect to x, remembering that $\dfrac{d}{dx}(f(y)) = \dfrac{d}{dy}f(y) \times \dfrac{dy}{dx}$.

Example

The point $P(2, 6)$ lies on the circle $x^2 + y^2 + 2x - 4y - 20 = 0$.

(a) Show that $\dfrac{dy}{dx} = \dfrac{x+1}{2-y}$

(b) Calculate the gradient of the tangent at P.

(a) $2x + 2y\dfrac{dy}{dx} + 2 - 4\dfrac{dy}{dx} + 0 = 0$

> Differentiate term by term with respect to x
> $\dfrac{d}{dx}(y) = \dfrac{dy}{dx}$.

> Divide throughout by 2 to simplify the equation.

$$x + y\frac{dy}{dx} + 1 - 2\frac{dy}{dx} = 0$$

$$x + 1 = 2\frac{dy}{dx} - y\frac{dy}{dx}$$

> Get the terms in $\dfrac{dy}{dx}$ on one side of the equation and all the other terms on the other side.

> Take out $\dfrac{dy}{dx}$ as a factor.

$$\Rightarrow x + 1 = \frac{dy}{dx}(2 - y)$$

$$\therefore \quad \frac{dy}{dx} = \frac{x+1}{2-y}$$

(b) At P, $x = 2$ and $y = 6$,

> The value of $\dfrac{dy}{dx}$ when $x = 2$ gives the gradient of the tangent at $(2, 6)$

so $\dfrac{dy}{dx} = \dfrac{3}{-4} = -\dfrac{3}{4}$.

The gradient of the tangent at P is $-\dfrac{3}{4}$.

Core 3 and Core 4 (Pure Mathematics)

Progress check

Core 3

1 Differentiate with respect to x:

(a) $y = 2 \sin 6x$ (b) $y = 4 \cos^3 x$ (c) $y = 2 \tan 3x$

(d) $y = e^{\sin x}$ (e) $y = \sin(x^3)$

Core 3

2 If $y = \sin^2 x$, find $\dfrac{dy}{dx}$

Core 3

3 Differentiate with respect to x:

(a) $\ln(x^2 + 3)$ (b) $\ln(2x - 4)^5$ (c) $\ln(\cos 2x)$ (d) $\ln(6x - 1)$

Core 3

4 Find $\dfrac{dy}{dx}$, simplifying your answers.

(a) $y = x^2 \cos 3x$ (b) $y = e^{2x} \sin x$ (c) $y = x \ln x$ (d) $y = (x + 1)(2x - 3)^4$

Core 3

5 Find the equation of the tangent to the curve $y = x\sqrt{1 + 2x}$ at the point (4, 12).

Core 3

6 Use the quotient rule to find $\dfrac{dy}{dx}$, simplifying your answers where necessary.

(a) $y = \dfrac{\sin 2x}{x}$ (b) $y = \dfrac{2x - 1}{3x + 4}$ (c) $y = \dfrac{\ln(x + 2)}{x + 2}$

Core 4

7 A curve is defined parametrically by $x = t^2$, $y = t^3$.

(a) Find $\dfrac{dy}{dx}$ when $t = 2$.

(b) Find the Cartesian equation of the curve.

Core 4

8 A curve has equation $x^2 + xy^2 - 5 = 0$.

(a) Find the value of $\dfrac{dy}{dx}$ at (1, 2).

(b) Find the equation of the normal when $x = 1$, in the form $ax + by + c = 0$.

1 (a) $12 \cos 6x$ (b) $-12 \sin x \cos^2 x$ (c) $6 \sec^2 3x$ (d) $\cos x e^{\sin x}$
(e) $3x^2 \cos(x^3)$
2 $2 \sin x \cos x$
3 (a) $\dfrac{2x}{x^2 + 3}$ (b) $\dfrac{5}{x - 2}$ (c) $-2 \tan 2x$ (d) $\dfrac{6}{6x - 1}$
4 (a) $x(2 \cos 3x - 3x \sin 3x)$ (b) $e^{2x}(\cos x + 2 \sin x)$
(c) $1 + \ln x$ (d) $5(2x + 1)(2x - 3)^3$
5 $3y = 13x - 16$
6 (a) $\dfrac{2x \cos 2x - \sin 2x}{x^2}$ (b) $\dfrac{11}{(3x + 4)^2}$ (c) $\dfrac{1 - \ln(x + 2)}{(x + 2)^2}$
7 (a) 3 (b) $x^3 = y^2$
8 (a) -1.5 (b) $2x - 3y + 4 = 0$

38

1.6 Integration

After studying this section you should be able to:

- integrate e^x and simple variations of this function with respect to x
- integrate trigonometrical functions
- recognise an integral that leads to a logarithmic function
- integrate using partial fractions
- integrate using a substitution
- use integration by parts to integrate a product
- use integration to find a volume of revolution
- solve first order differential equations when the variables can be separated
- understand exponential growth and decay

LEARNING SUMMARY

The exponential function e^x

Core 3

The fact that differentiation leaves the exponential function unchanged was stated on pages 31–32. Since integration is the reverse process of differentiation it follows that

> **KEY POINT**
>
> $$\int e^x \, dx = e^x + c \quad \text{and} \quad \int a e^x \, dx = a \int e^x \, dx = a e^x + c.$$

An extension of this result, using the reverse of the chain rule, is

> **KEY POINT**
>
> $$\int e^{ax+b} \, dx = \frac{1}{a} e^{ax+b} + c.$$

Examples

1 $\displaystyle\int 4 e^x \, dx = 4 e^x + c$ 2 $\displaystyle\int \tfrac{1}{2} e^{3x-1} \, dx = \tfrac{1}{6} e^{3x-1} + c$

Trigonometric functions

Core 3

The following integrals are obtained by applying the reverse process of differentiating the three main trigonometric functions (page 34).

$$\int \cos x \, dx = \sin x + c \qquad \int \cos(ax+b) \, dx = \frac{1}{a} \sin(ax+b) + c$$

$$\int \sin x \, dx = -\cos x + c \qquad \int \sin(ax+b) \, dx = -\frac{1}{a} \cos(ax+b) + c$$

$$\int \sec^2 x \, dx = \tan x + c \qquad \int \sec^2(ax+b) \, dx = \frac{1}{a} \tan(ax+b) + c$$

You should learn these and practise using them.

Example

It is a good idea to check your answer by differentiating.

$$\int (4 \cos 2x - \sin 2x) \, dx = 2 \sin 2x + \tfrac{1}{2} \cos 2x + c$$

Example

When integrating, the angle must be in radians.

Using exact values, this is

$2\left(1 - \dfrac{1}{\sqrt{3}}\right)$

$$\int_{\frac{\pi}{3}}^{\frac{\pi}{2}} \sec^2(\tfrac{1}{2}x)\,dx = \left[\frac{1}{\frac{1}{2}}\tan(\tfrac{1}{2}x)\right]_{\frac{\pi}{3}}^{\frac{\pi}{2}}$$

$$= 2\left[\tan\left(\tfrac{1}{2}x\right)\right]_{\frac{\pi}{3}}^{\frac{\pi}{2}}$$

$$= 2\left(\tan\frac{\pi}{4} - \tan\frac{\pi}{6}\right)$$

$$= 0.8452\ldots = 0.845 \ (3 \text{ s.f.})$$

These results are very useful and can be recognised easily:

$$\frac{d}{dx}(\sin^{n+1}x) = (n+1)\sin^n x \cos x \implies \int \sin^n x \cos x \, dx = \frac{1}{n+1}\sin^{n+1}x + c$$

$$\frac{d}{dx}(\cos^{n+1}x) = -(n+1)\cos^n x \sin x \implies \int \cos^n x \sin x \, dx = -\frac{1}{n+1}\cos^{n+1}x + c$$

Examples

$$\int \sin^5 x \cos x \, dx = \tfrac{1}{6}\sin^6 x + c$$

$$\int \sin x \cos^3 x \, dx = -\tfrac{1}{4}\cos^4 x + c$$

Odd powers of sin x and cos x

Example

Evaluate $\displaystyle\int_0^{\frac{\pi}{6}} \cos^3 x \, dx$

$$\int_0^{\frac{\pi}{6}} \cos^3 x \, dx = \int_0^{\frac{\pi}{6}} \cos x(\cos^2 x)\,dx$$

Use $\cos^2 x = 1 - \sin^2 x$.

$$= \int_0^{\frac{\pi}{6}} \cos x(1 - \sin^2 x)\,dx$$

$$= \int_0^{\frac{\pi}{6}} (\cos x - \cos x \sin^2 x)\,dx$$

Recognise the integral of $\cos x \sin^2 x$.

$$= \left[\sin x - \tfrac{1}{3}\sin^3 x\right]_0^{\frac{\pi}{6}}$$

$$= 0.458 \ (3 \text{ s.f.})$$

Core 4

Even powers of cos x and sin x

To integrate even powers of sin x and cos x, use the double angle identities for cos $2x$.

$$\cos 2x = 2\cos^2 x - 1 \implies \cos^2 x = \tfrac{1}{2}(1 + \cos 2x)$$

$$\cos 2x = 1 - 2\sin^2 x \implies \sin^2 x = \tfrac{1}{2}(1 - \cos 2x)$$

Example

$$\int \cos^2 x \; dx = \tfrac{1}{2} \int (1 + \cos 2x) dx$$

$$= \tfrac{1}{2}(x + \tfrac{1}{2} \sin 2x) + c$$

To integrate $\cos^4 x$, write it as $(\cos^2 x)^2 = (\tfrac{1}{2}(1 + \cos 2x))^2$.

The reciprocal function $\dfrac{1}{x}$

Core 3

The result $\displaystyle\int x^n \; dx = \dfrac{x^{n+1}}{n+1} + c$ holds for any value of n other than $n = -1$.

In this special case, the result takes a different form:

> Remember that $x^{-1} = \dfrac{1}{x}$.

$$\int \frac{1}{x} \; dx = \ln|x| + c.$$

KEY POINT

This extends to $\displaystyle\int \frac{k}{x} \; dx = k \int \frac{1}{x} \; dx = k \ln|x| + c,$

and to $\displaystyle\int \frac{1}{kx} \; dx = \frac{1}{k} \int \frac{1}{x} \; dx = \frac{1}{k} \ln|x| + c.$ (where k is a constant)

For example, $\displaystyle\int \frac{2}{x} \; dx = 2 \ln|x| + c$ and $\displaystyle\int \frac{1}{3x} \; dx = \frac{1}{3} \ln|x| + c.$

Integrals leading to a logarithmic function

Core 3

Remember that $\dfrac{d}{dx} \ln(f(x)) = \dfrac{f'(x)}{f(x)}.$

Applying the process in reverse gives:

$$\int \frac{f'(x)}{f(x)} \; dx = \ln |f(x)| + c.$$

KEY POINT

Examples

(a) $\displaystyle\int \frac{x^2}{2x^3 - 2} \; dx = \tfrac{1}{6} \ln | 2x^3 - 2 | + c$

> $\dfrac{d}{dx}(2x^3 - 2) = 6x^2$ so a factor of $\tfrac{1}{6}$ is needed.

(b) $\displaystyle\int \tan x \; dx = \int \frac{\sin x}{\cos x} \; dx$

$$= -\ln | \cos x | + c$$
$$= \ln | \sec x | + c$$

> $-\ln a = \ln(a^{-1}).$

Care must be taken with **definite integrals** of this type.
These can be evaluated between the limits a and b only if $f(x)$ exists for all values between a and b. Note that $f(a)$ and $f(b)$ will have the same sign.

$$\int_a^b \frac{f'(x)}{f(x)} \; dx = \left[\ln|f(x)| \right]_a^b = (\ln|f(b)| - \ln|f(a)|)$$

41

Key points from AS

• **Area under curve**
Revise AS page 33

Example

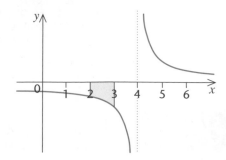

The diagram shows the curve $y = \dfrac{1}{x - 4}$.

Find the area between the curve, the x-axis, and the lines $x = 2$ and $x = 3$.

$$\int_a^b y \, dx = \int_2^3 \frac{1}{x - 4} \, dx$$

$$= \left[\ln|x - 4| \right]_2^3$$

$\ln|-2| = \ln 2$
$\ln 1 = 0$.

$$= \ln|-1| - \ln|-2|$$

$$= \ln 1 - \ln 2$$

$$= -\ln 2 \; (= -0.69 \ldots)$$

Note that the formula gives a negative value, confirming that the area is below the x-axis.

Shaded area = $\ln 2$ (= 0.69 ...) square units.

$f(3)$ and $f(5)$ do not have the same sign.

Note that $\displaystyle\int_3^5 \frac{1}{x - 4} \, dx$ cannot be evaluated, as the curve is undefined at $x = 4$.

Partial fractions

Core 4

Some rational functions can be integrated by expressing them in partial fraction form.

Example

Decompose into partial fractions (see page 15).

Find $\displaystyle\int \frac{x + 7}{(x - 2)(x + 1)} \, dx$, simplifying your answer.

$$\frac{x + 7}{(x - 2)(x + 1)} \equiv \frac{3}{x - 2} - \frac{2}{x + 1}$$

The working for this example is shown on page 15.

$$\Rightarrow \int \frac{x + 7}{(x - 2)(x + 1)} \, dx = \int \left(\frac{3}{x - 2} - \frac{2}{x + 1} \right) dx$$

$$= 3 \ln(x - 2) - 2 \ln(x + 1) + c$$

$$= \ln k \frac{(x - 2)^3}{(x + 1)^2}$$

Key points from AS

• **Logarithms**
Revise AS page 39

$\log a - \log b = \log\left(\dfrac{a}{b}\right)$

$\log a^n = n \log a$.

Note: Sometimes the constant c is incorporated into the log function by letting $c = \ln k$.

Substitution

Sometimes an integral is made easier by using a substitution.

This transforms the integral with respect to one variable, say x, into an integral with respect to a related variable, say u.

> **KEY POINT**
>
> $$\int f(x)\,\mathrm{d}x = \int f(x)\,\frac{\mathrm{d}x}{\mathrm{d}u}\,\mathrm{d}u.$$

This method is known as the method of **substitution** or **change of variable**.

> This could also be done by using the substitution $u = 2x + 1$.

Example

Find $\displaystyle\int x\sqrt{2x + 1}\,\mathrm{d}x$, using the substitution $u = \sqrt{2x + 1}$.

$$\int x\sqrt{2x + 1}\,\mathrm{d}x = \int x\sqrt{2x + 1}\,\frac{\mathrm{d}x}{\mathrm{d}u}\,\mathrm{d}u$$

Side working

> Show the side working as part of your answer.

$$= \int \tfrac{1}{2}(u^2 - 1)u \times u\,\mathrm{d}u$$

$u = \sqrt{2x + 1}$

$$= \frac{1}{2}\int (u^4 - u^2)\,\mathrm{d}u$$

$u^2 = 2x + 1 \Rightarrow x = \tfrac{1}{2}(u^2 - 1)$

> When simplifying, take out factors as soon as possible.

$$= \frac{1}{2}\left(\frac{u^5}{5} - \frac{u^3}{3}\right) + c$$

$\dfrac{\mathrm{d}x}{\mathrm{d}u} = u$

$$= \frac{u^3}{2}\left(\frac{u^2}{5} - \frac{1}{3}\right) + c$$

> Write the answer in terms of x, simplifying as much as possible.

$$= \tfrac{1}{30} u^3(3u^2 - 5) + c$$

$3u^2 - 5 = 3(2x + 1) - 5$

$$= \tfrac{1}{30}(2x + 1)^{\frac{3}{2}}(6x - 2) + c$$

$= 6x - 2$

$$= \tfrac{1}{15}(2x + 1)^{\frac{3}{2}}(3x - 1) + c$$

$= 2(3x - 1)$

When evaluating definite integrals using the method of substitution, change the x limits to u limits and substitute them into the working as soon as possible. There is no need to tidy up the expression first.

Sometimes it is possible to bypass the method of substitution by **recognising** an integral as the derivative of a particular function.

Examples

> This type of integral can always be done by using a substitution.

(a) $\displaystyle\int (2x + 1)^5\,\mathrm{d}x = \tfrac{1}{12}(2x + 1)^6 + c$

> Recognise that
> $$\frac{\mathrm{d}}{\mathrm{d}x}(2x + 1)^6 = 12(2x + 1)^5.$$

(b) $\displaystyle\int x\sqrt{2x^2 + 1}\,\mathrm{d}x = \tfrac{1}{6}(2x^2 + 1)^{\frac{3}{2}} + c$

> Recognise that
> $$\frac{\mathrm{d}}{\mathrm{d}x}(2x^2 + 1)^{\frac{3}{2}} = 6x(2x^2 + 1)^{\frac{1}{2}}.$$

General result:

$$\int f'(x)[f(x)]^n\,\mathrm{d}x = \frac{1}{n + 1}[f(x)]^{n+1} + c$$

Integration by parts

It is sometimes possible to integrate a product using **integration by parts**.

> **KEY POINT**
>
> If u and v are functions of x, then $\int u \dfrac{dv}{dx} dx = uv - \int v \dfrac{du}{dx} dx$.

Before using integration by parts, check whether the function can be simplified, or whether the integral can be recognised directly or done using a substitution.

Example

Find $\int 2x \sin x \, dx$.

Let $u = 2x$, then $\dfrac{du}{dx} = 2$

Let $\dfrac{dv}{dx} = \sin x$ then $v = -\cos x$

> Integration by parts can only be used if v can be found from $\dfrac{dv}{dx}$ and it is possible to find $\int v \dfrac{du}{dx} dx$.

$\therefore \quad \int 2x \sin x \, dx = 2x(-\cos x) - \int (-\cos x \times 2) dx$

$= -2x \cos x + \int 2 \cos x \, dx$

$= -2x \cos x + 2 \sin x + c$

Sometimes the integration by parts process has to be carried out more than once. This happens when integrating expressions such as $x^2 e^x$ or $x^2 \cos x$.

Definite integrals

$$\int_a^b u \frac{dv}{dx} dx = \left[uv \right]_a^b - \int_a^b v \frac{du}{dx} dx$$

Example

$\displaystyle\int_0^1 x e^{2x} \, dx = [x \tfrac{1}{2} e^{2x}]_0^1 - \int_0^1 \tfrac{1}{2} e^{2x} \, dx$

$= \tfrac{1}{2} e^2 - [\tfrac{1}{4} e^{2x}]_0^1$

$= \tfrac{1}{2} e^2 - (\tfrac{1}{4} e^2 - \tfrac{1}{4}) = \tfrac{1}{4} e^2 + \tfrac{1}{4}$

Working:

Let $u = x$, then $\dfrac{du}{dx} = 1$

Let $\dfrac{dv}{dx} = e^{2x}$, then $v = \tfrac{1}{2} e^{2x}$

Some functions containing $\ln x$ can be integrated using integration by parts. A useful strategy is to take $\ln x$ as u.

> Often a polynomial function is taken as u because it becomes of lower degree when differentiated. Integrals involving $\ln x$ are an exception, since $\ln x$ is not easy to integrate, but can be differentiated.

Example

$\displaystyle\int x^2 \ln x \, dx = \int (\ln x \times x^2) dx$

$= \ln x \times \left(\dfrac{x^3}{3}\right) - \int \left(\dfrac{x^3}{3} \times \dfrac{1}{x}\right) dx$

$= \tfrac{1}{3} x^3 \ln x - \tfrac{1}{3} \int x^2 \, dx$

$= \tfrac{1}{3} x^3 \ln x - \tfrac{1}{9} x^3 + c$

Working:

Let $u = \ln x$, then $\dfrac{du}{dx} = \dfrac{1}{x}$

Let $\dfrac{dv}{dx} = x^2$, then $v = \dfrac{x^3}{3}$

This method is particularly useful when integrating $\ln x$.
Think of it as integrating the product $\ln x \times 1$ as follows:

$$\int (\ln x \times 1)dx = \ln x \times x - \int \left(x \times \frac{1}{x}\right)dx$$

$$= x \ln x - \int 1 dx$$

$$= x \ln x - x + c$$

Working:

Let $u = \ln x$, then $\dfrac{du}{dx} = \dfrac{1}{x}$

Let $\dfrac{dv}{dx} = 1$, then $v = x$

Volume of revolution

Core 3

The shaded region R is bounded by the curve
$y = f(x)$, the x-axis and the lines $x = a$ and $x = b$.

Rotating R completely about the x-axis forms a
solid figure. The volume of this figure is called
the volume of revolution and is given by

$$V = \int_a^b \pi y^2 \, dx$$

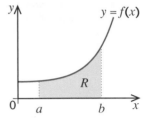

The corresponding result for rotation about the
y-axis is

$$\int_a^b \pi x^2 \, dy$$

where the region is bounded by the curve
$y = f(x)$, the y-axis and the lines $y - a$ and $y = b$.

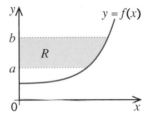

Example Find a formula for the volume of a cone of height h and base radius r.

The volume of the cone is given by the volume of
revolution of the shaded region shown.

The straight line has gradient $\dfrac{r}{h}$ and passes
through the origin, so its equation is $y = \dfrac{r}{h} x$.

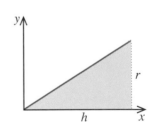

Using the formula for volume of revolution gives:

$$V = \int_0^h \pi \left(\frac{r}{h} x\right)^2 dx = \frac{\pi r^2}{h^2} \int_0^h x^2 \, dx$$

$$= \frac{\pi r^2}{h^2} \left[\frac{x^3}{3}\right]_0^h$$

$$= \frac{\pi r^2}{h^2} \times \frac{h^3}{3}$$

$$= \frac{\pi r^2 h}{3}.$$

Differential equations

Core 4

A **first order differential equation** in x and y contains only the first differential coefficient.

Examples $\dfrac{dy}{dx} = -2x,$ $\qquad \dfrac{dy}{dx} = \dfrac{x}{y},$ $\qquad y\dfrac{dy}{dx} = 3$

If $\dfrac{dy}{dx} = -2x,$ then integrating with respect to x gives $y = -x^2 + c.$

> The general solution contains an arbitrary integration constant.

This is the **general solution** of the differential equation and it can be illustrated by a family of curves.

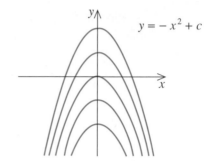

$$y = -x^2 + c$$

A **particular solution** is a specific member of the family and it can be found from additional information.

> The value of c is found to give a particular solution.

For example, if the curve $y = -x^2 + c$ goes through the point $(2, 0)$, then $0 = -(2)^2 + c \implies c = 4.$ The particular solution is $y = -x^2 + 4.$

Separating the variables

> **KEY POINT**
>
> A differential equation that can be written in the form $f(y)\dfrac{dy}{dx} = g(x)$
>
> can be solved by **separating the variables**, where $\displaystyle\int f(y)dy = \int g(x)dx.$

Example

Solve (a) $\dfrac{dy}{dx} = \dfrac{x}{y}$ $\qquad$ (b) $\dfrac{dy}{dx} = y$

(a) $\qquad \dfrac{dy}{dx} = \dfrac{x}{y}$

> Separate the variables to get the format $\int \dots dy = \int \dots dx.$

$$\int y\,dy = \int x\,dx \implies \tfrac{1}{2}y^2 = \tfrac{1}{2}x^2 + c$$

(b) $\qquad \dfrac{dy}{dx} = y$

> This is a special case, when $g(x) = 1.$

$$\frac{1}{y}\frac{dy}{dx} = 1$$

$$\int \frac{1}{y}dy = \int 1\,dx$$

> The answer can be left like this.

$$\ln y = x + c$$
$$y = e^{x+c} = e^x \times e^c = Ae^x \quad \text{(where } A = e^c\text{)}$$

So $\dfrac{dy}{dx} = y \implies y = Ae^x$

> The letter A is often used here, but it could be any letter (other than e, x or y, of course!).

Exponential growth and decay

Core 4

You may have to form a differential equation from given information. Often this models exponential growth or exponential decay.

Exponential growth

Exponential decay

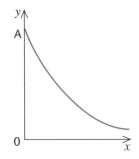

In exponential decay, the fact that y is decreasing with respect to x is usually shown by the negative, keeping $k > 0$.

$$\frac{dy}{dx} = ky \Rightarrow y = Ae^{kx} \quad (k > 0)$$

$$\frac{dy}{dx} = -ky \Rightarrow y = Ae^{-kx} \quad (k > 0)$$

An example of exponential growth is the growth of a population where

p_0 is the initial population.

$$\frac{dp}{dt} = kt \Rightarrow p = p_0 e^{-kt}$$

Examples of exponential decay are:

m_0 is the initial mass.

(a) the disintegration of radioactive materials, $\frac{dm}{dt} = -km \Rightarrow m = m_0 e^{-kt}$

T_s is temperature of surroundings, T_0 is initial temperature of the object.

(b) Newton's Law of Cooling, $\frac{dT}{dt} = -k(T - T_s) \Rightarrow T - T_s = (T_0 - T_s)e^{-kt}$

Progress check

Core 3

1 Integrate with respect to x.

(a) $\dfrac{3}{x}$ (b) $\dfrac{x+3}{x^2}$ (c) $3e^x - \dfrac{1}{x}$ (d) $3e^{4-x}$

Core 3

2 The region bounded by the curve $y = e^{\frac{x}{2}}$, the lines $x = 1$ and $x = 3$ and the x-axis is rotated completely about the x-axis. Find the volume of revolution.

Core 3

Parts (d) and (e) Core 4

3 (a) $\displaystyle\int 4 \sin 2x \, dx$ (b) $\displaystyle\int \sec^2 2x \, dx$ (c) $\displaystyle\int \cos\left(\tfrac{1}{2}x\right) dx$

(d) $\displaystyle\int 4 \sin^2 x \, dx$ (e) $\displaystyle\int \cos^2 2x \, dx$ (f) Evaluate $\displaystyle\int_0^{\frac{1}{6}\pi} \sin^4 x \cos x \, dx$

Core 3

4 Find (a) $\displaystyle\int \dfrac{1}{2x+5} \, dx$

(b) $\displaystyle\int \cot x \, dx$

Part (c) Core 4

(c) $\displaystyle\int \dfrac{3x+2}{x^2 - x - 12} \, dx$

> Split into partial fractions first.

Core 3

5 Using the substitution $u = 2 + e^{2x}$, or otherwise, find $\displaystyle\int \dfrac{e^{2x}}{2 + e^{2x}} \, dx$

> Remember that $\dfrac{dx}{du} = \dfrac{1}{du/dx}$.

Core 3

6 (a) Find (i) $\displaystyle\int xe^{-x} \, dx$ (ii) $\displaystyle\int x^3 \ln x \, dx$ (b) Evaluate $\displaystyle\int_0^{\frac{1}{2}\pi} x \cos x \, dx$

Core 4

7 Solve the differential equation $\dfrac{dy}{dx} = 2xy$, given that $y = 3$ when $x = 0$.

---

1 (a) $3 \ln|x| + c$ (b) $\ln|x| + c$ (c) $3e^x - \ln|x| + c$ (d) $-3e^{4-x} + c$

2 $\displaystyle\int_1^3 \pi e^x \, dx = \pi[e^x]_1^3 = \pi(e^3 - e)$.

3 (a) $-2 \cos 2x + c$ (b) $\tfrac{1}{2} \tan 2x + c$ (c) $2 \sin\left(\tfrac{1}{2}x\right) + c$
(d) $2x - \sin 2x + c$ (e) $\tfrac{x}{2} + \tfrac{1}{8} \sin 4x + c$ (f) $\tfrac{1}{160}$

4 (a) $\tfrac{1}{2} \ln(2x+5) + c$ (b) $\ln \sin(x) + c$ (c) $\ln(x+3)(x-4)^2 + c$ or $\ln A(x+3)(x-4)^2$

5 $\tfrac{1}{2} \ln(2 + e^{2x}) + c$

6 (a) (i) $-e^{-x}(x+1) + c$ (ii) $\tfrac{1}{16} x^4 (4 \ln x - 1) + c$ (b) $\tfrac{1}{2}\pi - 1$

7 $y = 3e^{x^2}$

1.7 Numerical methods

After studying this section you should be able to:

- *locate an interval containing a root of an equation by finding a change of sign*
- *use simple iteration to solve equations*
- *find the approximate value of an integral using Simpson's rule and the mid-ordinate rule*

LEARNING SUMMARY

Change of sign

Core 3

If there is an interval from $x = a$ to $x = b$ in which the graph of a function has no breaks then the function is said to be **continuous** on the interval.

All polynomial functions are continuous everywhere.

For example, the graph of $y = x^2 - 3$ is continuous everywhere.

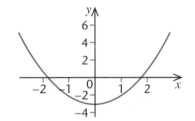

However, the graph of $y = \dfrac{1}{x - 3}$ is not continuous on any interval that contains the value $x = 3$.

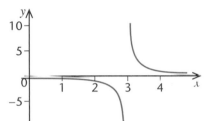

> **KEY POINT**
>
> If $f(x)$ is continuous between $x = a$ and $x = b$ and if $f(a)$ and $f(b)$ have different signs, then a root of $f(x) = 0$ lies in the interval from a to b.

Example Given that $f(x) = e^x - 10x$, show that the equation $f(x) = 0$ has a root between 3 and 4.

$f(3) = -9.914\ldots < 0$

$f(4) = 14.598\ldots > 0.$

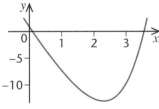

The change of sign shows that $f(x) = 0$ has a root between 3 and 4.

You can continue this process, called a **decimal search**, to trap the root in a smaller and smaller interval. The table gives the values of the function in steps of 0.1.

Most graphic calculators can tabulate values of a function and this gives a very efficient way to obtain the information.

x	3.1	3.2	3.3	3.4	3.5	3.6	3.7	3.8	3.9
$f(x)$	−8.802	−7.467	−5.887	−4.035	−1.884	0.5982	3.4473	6.7011	10.402

This shows that the root lies between 3.5 and 3.6
At the mid-point $f(3.55) = -0.6866\ldots < 0$ and so the root lies between 3.55 and 3.6.

This is a good way to establish the value of a root to a particular level of accuracy.

| 3.5 | 3.55 | 3.6 |

It follows that the value of the root is 3.6 to 1 d.p.

Iteration

Core 3

To solve an equation by **iteration** you start with some approximation to a root and improve its accuracy by substituting it into a formula. You can then repeat the process until you have the desired level of accuracy.

There are different methods for producing an iterative formula. You only need to know about simple iteration for this component.

Using **simple iteration**, the first step is to rearrange the equation to express x as a function of itself. This function defines the iterative formula that you need.

For example, the equation $e^x - 10x = 0$ can be rearranged as

$$e^x = 10x$$
$$\Rightarrow x = \ln(10x).$$

This may now be turned into an iterative formula for finding x.

$$x_{n+1} = \ln(10x_n).$$

Starting with $x_1 = 3$, $x_2 = \ln 30 = 3.401 \ldots$ and so on. This produces a sequence of values:

The values may be found very easily using a calculator's Ans function:
Key in 3 = followed by
ln(10Ans) = = =
to produce the sequence.
(Depending on the make of your calculator you may need to use EXE or ENTER in place of = .)

$x_2 = 3.401 \ldots$	$x_6 = 3.5760 \ldots$
$x_3 = 3.526 \ldots$	$x_7 = 3.5768 \ldots$
$x_4 = 3.563 \ldots$	$x_8 = 3.5770 \ldots$
$x_5 = 3.573 \ldots$	$x_9 = 3.5771 \ldots$

The calculator display soon settles on 3.577152064. The value of the root is $x = 3.577$ to 4 s.f.

The diagram shows how the process converges from the starting point on either side of the root.

Start from a point on the x-axis, move up to the curve and across to $y = x$. Then move up to the curve and across to $y = x$ again. Continue in the same way. Each of these stages corresponds to one iteration of the formula.

A different rearrangement of the original equation gives $x = \dfrac{e^x}{10}$ and so the corresponding iterative

formula is $x_{n+1} = \dfrac{e^{x_n}}{10}$.

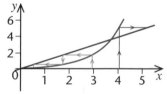

You may need to try several rearrangements in order to converge to a particular root of an equation by this method.

Starting with $x = 3$, this rearrangement fails to converge to the root between 3 and 4 but it does converge to the other root of the equation.

Notice how the movement is away from the upper root this time. If the starting value is smaller than the upper root then the process converges to the lower root.
A starting value above the upper root fails to converge to either root.

The value of this root is $x = 0.1118$ to 4 s.f.

In some exam questions you may be given the iteration formula to start with.

Example Starting with $x_1 = 1$, use the formula $x_{n+1} = \dfrac{x_n^2 - 5x_n}{10} - 3$ to find $x_2, x_3, \ldots, x_8$ and describe the long term behaviour of the sequence.

Try the key sequence:
1 =
$(Ans^2 - 5Ans)/10 - 3$
=
=
=
=

The sequence oscillates but converges to −1.787 to 4 s.f. which is the lower root of the equation

$$x = \frac{x^3 - 5x}{10} - 3,$$

which simplifies to

$$x^2 - 15x - 30 = 0.$$

The oscillations correspond to a cobweb pattern on the diagram.

Numerical integration

Core 3

Simpson's rule

This method divides the area into an **even number** of parallel strips and approximates the areas of pairs of strips using the following formula:

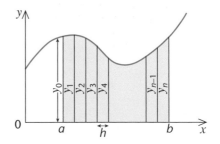

Note that the first ordinate has been labelled y_0 and the last y_n. There are n strips and $n+1$ ordinates.

KEY POINT

$$\int_a^b y\,dx \approx \tfrac{1}{3}h\{(y_0 + y_n) + 4(y_1 + y_3 + ... + y_{n-1}) + 2(y_2 + y_4 + ... + y_{n-2})\}$$

where $h = \dfrac{b-a}{n}$ and n is even.

Example

Estimate $\displaystyle\int_0^1 e^{x^2}\,dx$ using Simpson's rule with 11 ordinates (10 strips).

Tabulating the results helps in doing the final calculation.

If you have multiple memories in your calculator, retain the figures. Otherwise approximate say to 4 decimal places.

x	y	First and last ordinates	Odd ordinates	Other ordinates
0	y_0	1		
0.1	y_1		1.010 ...	
0.2	y_2			1.040 ...
0.3	y_3		1.094 ...	
0.4	y_4			1.173 ...
0.5	y_5		1.284 ...	
0.6	y_6			1.433 ...
0.7	y_7		1.632 ...	
0.8	y_8			1.896 ...
0.9	y_9		2.247 ...	
1	y_{10}	2.718 ...		
Totals		3.718 ...	7.268 ...	5.544 ...

$h = \dfrac{1-0}{10} = 0.1$

$$\int_a^b y\,dx \approx \tfrac{1}{3}h\{(y_0 + y_{10}) + 4(y_1 + y_3 + y_5 + y_7 + y_9) + 2(y_2 + y_4 + y_6 + y_8)\}$$

$$= \tfrac{1}{3} \times 0.1 \times \{3.718 ... + 4(7.268 ...) + 2(5.544 ...)\}$$

$$= 1.463 \text{ (3 s.f.)}$$

Core 3

The mid-ordinate rule

The area is split into rectangles of equal width h.

The height of each rectangle is taken as the y-value halfway along each rectangle.

> The more rectangles that are taken, the better the approximation.

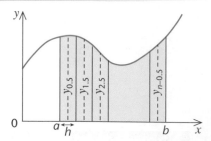

Denoting the mid-ordinates by $y_{0.5}$, $y_{1.5}$, ..., $y_{n-0.5}$, then

> **KEY POINT**
>
> $$\int_a^b y\,dx \approx h(y_{0.5} + y_{1.5} + \dots + y_{n-0.5}) \text{ where } h = \frac{b-a}{n}.$$

Example

Estimate $\int_1^3 \ln x\,dx$ using the mid-ordinate rule with 4 strips.

There are 4 strips, so $h = \dfrac{3-1}{4} = 0.5$,

$x_0 = 1$, $x_1 = 1.5$, $x_2 = 2$, $x_3 = 2.5$, $x_4 = 3$.

$\therefore \quad x_{0.5} = 1.25 \Rightarrow y_{0.5} = \ln 1.25$

Calculating the other y mid-ordinates in a similar way gives:

> Using integration by parts, the exact answer is $3\ln 3 - 2 = 1.295\ldots$ so the answers agree to 2 d.p.

$$\int_1^3 \ln y\,dx \approx 0.5(\ln 1.25 + \ln 1.75 + \ln 2.25 + \ln 2.75)$$

$$= 1.3026\ldots = 1.30 \text{ (2 d.p.)}$$

Progress check

Core 3

1 Show that the equation $x^3 - x - 7 = 0$ has a root between 2 and 3.
   Use a decimal search to find the value of the root to 2 d.p.

Core 3

2 Use the iteration $x_{n+1} = \sqrt[3]{x_n + 7}$ to refine the value of the root found in question 1 and state its value to 4 d.p.

Core 3

3 Estimate $\int_0^1 \sqrt{1 - x^3}\,dx$

   (a) using Simpson's rule with 5 ordinates (4 strips)
   (b) using the mid-ordinate rule with 4 strips.

3 To 3 d.p. (a) 0.823 (b) 0.854
2 2.0867
1 2.09

1.8 Vectors

After studying this section you should be able to:

- *understand the distinction between vector and scalar quantities*
- *add and subtract vector quantities and multiply by a scalar*
- *understand the vectors* **i**, **j** *and* **k**
- *find the magnitude and direction of a vector*
- *use scalar products to find the angle between two directions*
- *express equations of lines in vector form*
- *determine whether two lines are parallel, intersect or are skew*

Vector and scalar quantities

Core 4

A **scalar** quantity has size (or **magnitude**) but not direction. **Numbers** are scalars and some other important examples are **distance**, **speed**, **mass** and **time**.

A **vector** quantity has both size and **direction**. For example, distance in a specified direction is called **displacement**.

The diagram shows a **directed line segment**. It has size (in this case, length) and direction so it is a vector.

The diagram gives a useful way to represent *any* vector quantity and may also be used to represent addition and subtraction of vectors and multiplication of a vector by a scalar.

In a textbook, vectors are usually labelled with lower case letters in **bold** print.
When hand-written, these letters should be underlined e.g. *a*.

Addition and subtraction of vectors

Core 4

This diagram shows three vectors **a**, **b** and **c** such that **c** = **a** + **b**.

The vectors **a** and **b** follow on from each other and then **c** joins the start of vector **a** to the end of vector **b**.

Check that **a** + **b** = **b** + **a**.

The vector **c** is called the **resultant** of **a** and **b**.

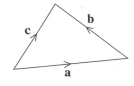

On a vector diagram, −**q** has the opposite sense of direction to **q**.
Notice that **p** − **q** is represented as **p** + (−**q**).

This diagram shows the same information in a different way.

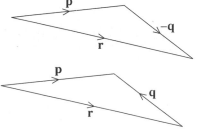

Following the route in the opposite direction to **q** is the same as adding −**q** or subtracting **q**.

Scalar multiplication

Core 4

2p is parallel to **p** and has the same sense of direction but is twice as long.

−3p is parallel to **p** but has the opposite sense of direction and is three times as long.

Component form

Core 4

A unit vector is a vector of magnitude 1 unit.

i, **j**, **k** or column format may be used.

When working in two dimensions, it is often very useful to express a vector in terms of two special vectors **i** and **j**. These are **unit vectors** at right-angles to each other.

A vector **r** written as $\mathbf{r} = a\mathbf{i} + b\mathbf{j}$ is said to have **components** $a\mathbf{i} + b\mathbf{j} = \begin{bmatrix} a \\ b \end{bmatrix}$

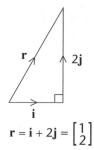

$$\mathbf{r} = \mathbf{i} + 2\mathbf{j} = \begin{bmatrix} 1 \\ 2 \end{bmatrix}$$

$$\mathbf{r} = 3\mathbf{i} - \mathbf{j} = \begin{bmatrix} 3 \\ -1 \end{bmatrix}$$

Examples

For work involving the Cartesian coordinate system, **i** and **j** are taken to be in the positive directions of the *x*- and *y*-axes respectively.

In three dimensions, a third vector **k** is used to represent a unit vector in the positive direction of the *z*-axis.

The **position vector** of a point P is the vector $\overrightarrow{OP}$ where O is the origin.

If P has coordinates (a, b) then its position vector is given by $\mathbf{r} = a\mathbf{i} + b\mathbf{j} = \begin{bmatrix} a \\ b \end{bmatrix}$

In three dimensions, a point with coordinates (a, b, c) would have position vector $\mathbf{r} = a\mathbf{i} + b\mathbf{j} + c\mathbf{k} = \begin{bmatrix} a \\ b \\ c \end{bmatrix}$

Adding and subtracting vectors in component form

Core 4

Addition and subtraction may be done algebraically.

Example

$\mathbf{p} = 3\mathbf{i} + 2\mathbf{j} - \mathbf{k}$ and $\mathbf{q} = 5\mathbf{i} - \mathbf{j} + 4\mathbf{k}$. Express the following vectors in terms of **i**, **j** and **k**:

(a) $\mathbf{p} + \mathbf{q}$ (b) $\mathbf{p} - \mathbf{q}$ (c) $2\mathbf{p} - 3\mathbf{q}$.

$\mathbf{p} + \mathbf{q} = \begin{bmatrix} 3 \\ 2 \\ -1 \end{bmatrix} + \begin{bmatrix} 5 \\ -1 \\ 4 \end{bmatrix} = \begin{bmatrix} 8 \\ 1 \\ 3 \end{bmatrix}$

(a) $\mathbf{p} + \mathbf{q} = (3\mathbf{i} + 2\mathbf{j} - \mathbf{k}) + (5\mathbf{i} - \mathbf{j} + 4\mathbf{k}) = 8\mathbf{i} + \mathbf{j} + 3\mathbf{k}$

(b) $\mathbf{p} - \mathbf{q} = (3\mathbf{i} + 2\mathbf{j} - \mathbf{k}) - (5\mathbf{i} - \mathbf{j} + 4\mathbf{k}) = -2\mathbf{i} + 3\mathbf{j} - 5\mathbf{k}$

(c) $2\mathbf{p} - 3\mathbf{q} = 2(3\mathbf{i} + 2\mathbf{j} - \mathbf{k}) - 3(5\mathbf{i} - \mathbf{j} + 4\mathbf{k})$
$= 6\mathbf{i} + 4\mathbf{j} - 2\mathbf{k} - 15\mathbf{i} + 3\mathbf{j} - 12\mathbf{k} = -9\mathbf{i} + 7\mathbf{j} - 14\mathbf{k}$.

Finding the magnitude of a vector

Core 4

$\mathbf{r} = \begin{bmatrix} a \\ b \\ c \end{bmatrix}$

KEY POINT

If $\mathbf{r} = a\mathbf{i} + b\mathbf{j} + c\mathbf{k}$, the magnitude (length) of $\mathbf{r}$ is given by

$$|\mathbf{r}| = \sqrt{a^2 + b^2 + c^2}$$

$\mathbf{r} = \begin{bmatrix} 3 \\ -4 \\ 12 \end{bmatrix}$

For example,
If $\mathbf{r} = 3\mathbf{i} - 4\mathbf{j} + 12\mathbf{k}$, then $|\mathbf{r}| = \sqrt{3^2 + (-4)^2 + 12^2} = 13$.

Scalar product

Core 4

If $\mathbf{a} = x_1\mathbf{i} + y_1\mathbf{j} + z_1\mathbf{k}$, then $|\mathbf{a}| = \sqrt{x_1^2 + y_1^2 + z_1^2}$ where $|\mathbf{a}|$ is the length of $\mathbf{a}$.

If $\mathbf{b} = x_2\mathbf{i} + y_2\mathbf{j} + z_2\mathbf{k}$, then $|\mathbf{b}| = \sqrt{x_2^2 + y_2^2 + z_2^2}$.

KEY POINT

If θ is the angle between $\mathbf{a}$ and $\mathbf{b}$,
the **scalar (or dot) product** of $\mathbf{a}$ and $\mathbf{b}$ is $\mathbf{a.b}$
where $\mathbf{a.b} = |\mathbf{a}||\mathbf{b}| \cos \theta$.

This leads to the useful results:
$\mathbf{i.i} = 1$, $\mathbf{j.j} = 1$, $\mathbf{k.k} = 1$
$\mathbf{i.j} = \mathbf{j.k} = \mathbf{k.i} = 0$.

If two vectors are **parallel**, $\theta = 0$ and $\mathbf{a.b} = |\mathbf{a}||\mathbf{b}|$.
If two vectors are **perpendicular**, $\theta = 90°$ and $\mathbf{a.b} - 0$.

For $\mathbf{a}$ and $\mathbf{b}$ defined as above:

$\mathbf{a.b} = \begin{bmatrix} x_1 \\ y_1 \\ z_1 \end{bmatrix} . \begin{bmatrix} x_2 \\ y_2 \\ z_2 \end{bmatrix}$

$= x_1 x_2 + y_1 y_2 + z_1 z_2$.

$$\mathbf{a.b} = (x_1\mathbf{i} + y_1\mathbf{j} + z_1\mathbf{k}) . (x_2\mathbf{i} + y_2\mathbf{j} + z_2\mathbf{k})$$
$$= x_1 x_2 + y_1 y_2 + z_1 z_2$$

This uses the results for parallel and perpendicular vectors.

Example
Find the angle between $\mathbf{a} = 2\mathbf{i} + \mathbf{j} + 4\mathbf{k}$ and $\mathbf{b} = -3\mathbf{i} + 2\mathbf{j} - \mathbf{k}$.

$$\mathbf{a.b} = 2(-3) + 1(2) + 4(-1) = -8$$
$$|\mathbf{a}| = \sqrt{2^2 + 1^2 + 4^2} = \sqrt{21}$$
$$|\mathbf{b}| = \sqrt{(-3)^2 + 2^2 + (-1)^2} = \sqrt{14}$$
$$\mathbf{a.b} = |\mathbf{a}||\mathbf{b}| \cos \theta$$
$$-8 = \sqrt{21}\sqrt{14} \cos \theta$$
$$\Rightarrow \cos \theta = \frac{-8}{\sqrt{21}\sqrt{14}} = -0.4665 \dots$$
$$\Rightarrow \theta = 117.8° \text{ (1 d.p.)}$$

$\mathbf{a.b} = \begin{bmatrix} 2 \\ 1 \\ 4 \end{bmatrix} . \begin{bmatrix} -3 \\ 2 \\ -1 \end{bmatrix} = -8.$

Vector equation of a line

Core 4

KEY POINT

A **vector equation** of a line passing
through a fixed point A with position
vector $\mathbf{a}$ and parallel to a vector $\mathbf{b}$ is
$\mathbf{r} = \mathbf{a} + t\mathbf{b}$, where t is a scalar
parameter.

t is often used for the
parameter, but not
exclusively so. Other
letters sometimes used
include s, μ and λ.

r is the position vector of any point on the line.
b is the **direction vector** of the line.

The equation of the line is not unique. Any other point on the line could be used instead of A and any multiple of **b** could be used for the direction vector.

Example

In column format:

$$r = \begin{bmatrix} 4 \\ -5 \\ 1 \end{bmatrix} = t\begin{bmatrix} 3 \\ 4 \\ -2 \end{bmatrix}$$

A vector equation of the line passing through $(4, -5, 1)$, parallel to $3\mathbf{i} + 4\mathbf{j} - 2\mathbf{k}$ is
$\mathbf{r} = 4\mathbf{i} - 5\mathbf{j} + \mathbf{k} + t(3\mathbf{i} + 4\mathbf{j} - 2\mathbf{k})$.

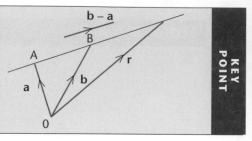

KEY POINT

A vector equation of the line through two fixed points A and B, with position vectors **a** and **b**, is given by
$\mathbf{r} = \mathbf{a} + t(\mathbf{b} - \mathbf{a})$ where t is a scalar parameter.

Example

Find a vector equation of the line through the points $P(3, 1, -4)$ and $Q(-2, 5, 1)$.

Direction of line PQ:

$$q - p = \begin{bmatrix} -2 \\ 5 \\ 1 \end{bmatrix} - \begin{bmatrix} 3 \\ 1 \\ -4 \end{bmatrix} = \begin{bmatrix} -5 \\ 4 \\ 5 \end{bmatrix}$$

$\overrightarrow{PQ} = \mathbf{q} - \mathbf{p} = -2\mathbf{i} + 5\mathbf{j} + \mathbf{k} - (3\mathbf{i} + \mathbf{j} - 4\mathbf{k}) = -5\mathbf{i} + 4\mathbf{j} + 5\mathbf{k}$

A vector equation of PQ is $\mathbf{r} = 3\mathbf{i} + \mathbf{j} - 4\mathbf{k} + t(-5\mathbf{i} + 4\mathbf{j} + 5\mathbf{k})$.

$$r = \begin{bmatrix} 3 \\ 1 \\ -4 \end{bmatrix} + t\begin{bmatrix} -5 \\ 4 \\ 5 \end{bmatrix}$$

Angles between two lines

To find the angle between two lines, find the angle between their direction vectors.

Pairs of lines

Core 4

In 2-dimensions, a pair of lines are either parallel or they intersect.
In 3-dimensions, a pair of lines are parallel, or they intersect or they are **skew**.

If two lines are parallel, then their direction vectors are multiples of each other, as in this **example**:
$\mathbf{r} = 3\mathbf{i} - 5\mathbf{j} + 2\mathbf{k} + \lambda(\mathbf{i} + 2\mathbf{j} - \mathbf{k})$ and $\mathbf{r} = 4\mathbf{i} + \mathbf{j} - 2\mathbf{k} + \mu(3\mathbf{i} + 6\mathbf{j} - 3\mathbf{k})$.

$$r = \begin{bmatrix} 3 \\ -5 \\ 2 \end{bmatrix} + \lambda\begin{bmatrix} 1 \\ 2 \\ -1 \end{bmatrix}$$

$$r = \begin{bmatrix} 4 \\ 1 \\ -2 \end{bmatrix} + \mu\begin{bmatrix} 3 \\ 6 \\ -3 \end{bmatrix}$$

The two lines $\mathbf{r} = \mathbf{a} + \lambda\mathbf{b}$ and $\mathbf{r} = \mathbf{c} + \mu\mathbf{d}$ intersect if unique values of λ and μ can be found such that $\mathbf{a} + \lambda\mathbf{b} = \mathbf{c} + \mu\mathbf{d}$.
If unique values cannot be found, then the two lines are **skew**.

Example

If the lines intersect, then $\mathbf{r}_1 = \mathbf{r}_2$ will have a unique solution for λ and μ.

Note that the lines are not parallel, since their direction vectors are not multiples of each other.

Investigate whether the lines $\mathbf{r}_1 = \begin{bmatrix} 2 \\ 1 \\ -3 \end{bmatrix} + \lambda\begin{bmatrix} 4 \\ 6 \\ -1 \end{bmatrix}$ and $\mathbf{r}_2 = \begin{bmatrix} 3 \\ 0 \\ -2 \end{bmatrix} + \mu\begin{bmatrix} 4 \\ 1 \\ 3 \end{bmatrix}$ intersect or whether they are skew.

$$\mathbf{r}_1 = \mathbf{r}_2 \Rightarrow \begin{bmatrix} 2 \\ 1 \\ -3 \end{bmatrix} + \lambda\begin{bmatrix} 4 \\ 6 \\ -1 \end{bmatrix} = \begin{bmatrix} 3 \\ 0 \\ -2 \end{bmatrix} + \mu\begin{bmatrix} 4 \\ 1 \\ 3 \end{bmatrix}$$

i.e. $\begin{bmatrix} 2+4\lambda \\ 1+6\lambda \\ -3-\lambda \end{bmatrix} = \begin{bmatrix} 3+4\mu \\ 0+\mu \\ -2+3\mu \end{bmatrix}$

Equating coefficients:

Solving (1) and (2) simultaneously gives:

$$2 + 4\lambda = 3 + 4\mu \quad (1) \qquad \lambda = -0.25 \text{ and } \mu = -0.5$$
$$1 + 6\lambda = \mu \quad (2) \qquad \text{Check whether these satisfy (3):}$$
$$-3 - \lambda = -2 + 3\mu \quad (3) \qquad \text{LHS} = -3 - (-0.25) = -2.75$$
$$\text{RHS} = -2 + 3(-0.5) = -3.5$$

$\therefore \lambda = -0.25$ and $\mu = -0.5$ do not satisfy all three equations.

Since there is not a unique solution for λ and μ, the lines do not intersect. They are skew.

Progress check

Core 4

1 Write down the position vector **r** of a point with coordinates (5, −2).

Core 4

2 Find the magnitude of the vector **r** = 5**i** − 12**j** and give its direction relative to **i**.

Core 4

3 Find the angle between the vectors $\mathbf{a} = 2\mathbf{i} + 3\mathbf{j} - \mathbf{k} = \begin{bmatrix} 2 \\ 3 \\ -1 \end{bmatrix}$ and $\mathbf{b} = \mathbf{i} + \mathbf{j} + 2\mathbf{k} = \begin{bmatrix} 1 \\ 1 \\ 2 \end{bmatrix}$.

Core 4

4 (a) Give a vector equation of the line which is parallel to $\begin{bmatrix} 4 \\ -1 \\ 2 \end{bmatrix}$ and goes through the point with coordinates (3, 0, 1).

(b) Give a vector equation of the line through (4, −1, 1) and (−3, 2, −2).

Core 4

5 (a) Find the point of intersection of the lines

$$\mathbf{r} = \begin{bmatrix} 2 \\ -3 \end{bmatrix} + \mu \begin{bmatrix} 1 \\ -2 \end{bmatrix} \text{ and } \mathbf{r} = \begin{bmatrix} 1 \\ -2 \end{bmatrix} + \lambda \begin{bmatrix} 2 \\ 3 \end{bmatrix}.$$

(b) Investigate whether the following lines intersect or whether they are skew. If they intersect, find their point of intersection.

$$\mathbf{r} = \begin{bmatrix} 1 \\ 2 \\ -5 \end{bmatrix} + s \begin{bmatrix} 3 \\ 1 \\ 1 \end{bmatrix} \text{ and } \mathbf{r} = \begin{bmatrix} 2 \\ -1 \\ 2 \end{bmatrix} + t \begin{bmatrix} -1 \\ -1 \\ 1 \end{bmatrix}.$$

(c) Find the angle between the lines given in (b).

5 (a) $(1\frac{2}{3}, -1\frac{1}{3})$ (b) Intersect at (7, 4, −3) (c) 121.5°

(b) $\mathbf{r} = \begin{bmatrix} 4 \\ -1 \\ 1 \end{bmatrix} + t \begin{bmatrix} 3 \\ 3 \\ 3 \end{bmatrix}$, i.e. $\mathbf{r} = 4\mathbf{i} - \mathbf{j} + \mathbf{k} + t(-7\mathbf{i} + 3\mathbf{j} - 3\mathbf{k})$

(a) $\mathbf{r} = \begin{bmatrix} 3 \\ 0 \\ 1 \end{bmatrix} + t \begin{bmatrix} 4 \\ -1 \\ 2 \end{bmatrix}$, i.e. $\mathbf{r} = 3\mathbf{i} + \mathbf{k} + t(4\mathbf{i} - \mathbf{j} + 2\mathbf{k})$

4 (Answers are not unique, other answers are possible.)

3 70.9°

2 $|\mathbf{r}| = 13$, 67.4° clockwise.

1 $\mathbf{r} = 5\mathbf{i} - 2\mathbf{j} = \begin{bmatrix} 5 \\ -2 \end{bmatrix}$

Sample questions and model answers

Core 4

1

(a)   Express $f(x) = \dfrac{5x+1}{(x+2)(x-1)^2}$ in partial fractions.

(b)   Hence show that $\displaystyle\int_{-1}^{0} f(x)\,dx = 1 - 2\ln 2$.

Remember the format when there is a repeated factor.

Take care with the denominator.

(a)   Let $\dfrac{5x+1}{(x+2)(x-1)^2} \equiv \dfrac{A}{x+2} + \dfrac{B}{x-1} + \dfrac{C}{(x-1)^2}$

$\Rightarrow \dfrac{5x+1}{(x+2)(x-1)^2} \equiv \dfrac{A(x-1)^2 + B(x+2)(x-1) + C(x+2)}{(x+2)(x-1)^2}$

$\Rightarrow 5x+1 \equiv A(x-1)^2 + B(x+2)(x-1) + C(x+2)$

Putting $x = 1$ makes the factor $(x-1)$ equal to zero.

You could equate the x term or the constant term if you wish.

Let $x = 1$: $\qquad 6 = 3C \Rightarrow C = 2$

Let $x = -2$: $\qquad -9 = 9A \Rightarrow A = -1$

Equate x^2 terms: $\quad 0 = A + B \Rightarrow B = 1$

$\therefore\; f(x) = -\dfrac{1}{x+2} + \dfrac{1}{x-1} + \dfrac{2}{(x-1)^2}$

(b) $\displaystyle\int_{-1}^{0} f(x)\,dx = \int_{-1}^{0}\left(-\dfrac{1}{x+2} + \dfrac{1}{x-1} + \dfrac{2}{(x-1)^2}\right)dx$

Remember the modulus sign.

$= \left[-\ln|x+2| + \ln|x-1| - 2(x-1)^{-1}\right]_{-1}^{0}$

$= -\ln 2 + \ln|-1| + 2 - (-\ln 1 + \ln|-2| + 1)$

$= -\ln 2 + 2 - \ln 2 - 1$

$\ln 1 = \ln|-1| = 0$
$\ln|-2| = \ln 2$.

$= 1 - 2\ln 2$

Core 4

2

Expand $(1 - 2x)^{-3}$ in ascending powers of x as far as the term in x^3 and state the set of values of x for which the expansion is valid.

Take care with the negatives.

Remember to include the restriction on x, even when it is not requested specifically in the question.

$(1-2x)^{-3} = 1 + (-3)(-2x) + \dfrac{(-3)(-4)}{2!}(-2x)^2 + \dfrac{(-3)(-4)(-5)}{3!}(-2x)^3 + \ldots$

$= 1 + 6x + 24x^2 + 80x^3 + \ldots \qquad$ provided $|2x| < 1$, i.e. $|x| < 0.5$

Sample questions and model answers (continued)

Core 4

3

Given $\tan \dfrac{\pi}{4} = 1$, use the substitution $x = \tan u$ to show that $\displaystyle\int_0^1 \dfrac{1}{(1+x^2)^2}\,dx = \int_0^{\frac{\pi}{4}} \cos^2 u\,du$.

Hence show that $\displaystyle\int_0^1 \dfrac{1}{(1+x^2)^2}\,dx = \dfrac{\pi}{8} + \dfrac{1}{4}$.

> Show this essential side working.

$$\int_0^1 \frac{1}{(1+x^2)^2}\,dx = \int_{x=0}^{x=1} \frac{1}{(1+x^2)^2}\frac{dx}{du}\,du \qquad \text{Let } x = \tan u$$

Remember to change the x limits to u limits.

$$= \int_0^{\frac{\pi}{4}} \frac{1}{(\sec^2 u)^2} \sec^2 u\,du \qquad \Longrightarrow \frac{dx}{du} = \sec^2 u$$

$$= \int_0^{\frac{\pi}{4}} \frac{1}{\sec^2 u}\,du \qquad 1 + x^2 = 1 + \tan^2 u$$

$$= \sec^2 u$$

To integrate even powers of sin x or cos x, change to double angles using cos 2x formula.

$$= \int_0^{\frac{\pi}{4}} \cos^2 u\,du \qquad \text{Limits:}$$

When $x = 0$, $\tan u = 0 \Longrightarrow u = 0$

When $x = 1$, $\tan u = 1 \Longrightarrow u = \dfrac{\pi}{4}$

$$\int_0^{\frac{\pi}{4}} \cos^2 u\,du = \frac{1}{2}\int_0^{\frac{\pi}{4}} (1 + \cos 2u)\,du \qquad \cos 2u = 2\cos^2 u - 1$$

$$= \frac{1}{2}\left[u + \frac{1}{2}\sin 2u \right]_0^{\frac{\pi}{4}} \qquad \Longrightarrow \cos^2 u = \tfrac{1}{2}(1 + \cos 2u)$$

$$= \frac{1}{2}\left(\frac{\pi}{4} + \frac{1}{2} - 0 \right)$$

$$= \frac{\pi}{8} + \frac{1}{4}$$

Core 3

4

(a) Solve the inequality $|2x - 3| < 11$

(b) Sketch the graph of $y = |f(x)|$ where $f(x) = (x-1)(x-2)(x-3)$

(a) $|2x - 3| < 11$

This first step is important.

$$\Longrightarrow -11 < 2x - 3 < 11$$

You can simplify the inequality by doing the same thing to each part.

$$\Longrightarrow -8 < 2x < 14 \qquad \text{Add 3 to each part.}$$

$$\Longrightarrow -4 < x < 7. \qquad \text{Divide each part by 2.}$$

(b)

Parts of the graph of $y = f(x)$ that would be below the x-axis are reflected in the x-axis to give the graph of $y = |f(x)|$.

Sample questions and model answers (continued)

Core 3

5

The functions f and g are defined by:

$$f(x) = \sqrt{x-1} \quad x \geqslant 1$$

$$g(x) = 2x + 3 \quad x \in \mathbb{R}$$

(a) Find an expression for $f^{-1}(x)$ and state its domain and range.

(b) Find the value of:

   (i) $fg(7)$

   (ii) $gf(10)$

   (iii) $f^{-1}g^{-1}(15)$

Set $y = f(x)$ and rearrange to make x the subject.

(a) $y = \sqrt{x-1}$

$$\Rightarrow y^2 = x - 1$$

$$\Rightarrow x = y^2 + 1$$

$$\Rightarrow f^{-1}(x) = x^2 + 1$$

The domain of f^{-1} is given by the range of f.

The domain of f^{-1} is $x \geqslant 0$ and the range of f^{-1} is $x \geqslant 1$.

The range of f^{-1} is given by the domain of f.

(b) (i) $fg(7) = f(17) = 4$

The order in which you apply the functions is important.

   (ii) $gf(10) = g(3) = 9$

   (iii) $f^{-1}g^{-1}(15) = f^{-1}(6) = 37$

Core 4

6

The equation of a straight line l is

$$\mathbf{r} = \begin{bmatrix} 4 \\ 0 \\ -2 \end{bmatrix} + t \begin{bmatrix} 3 \\ 0 \\ 1 \end{bmatrix} \quad \text{where } t \text{ is a scalar parameter.}$$

Points A, B and C lie on the line l. Referred to the origin O, they have position vectors $\mathbf{a}$, $\mathbf{b}$ and $\mathbf{c}$ respectively.

(a)   Find $\mathbf{a}$ and $\mathbf{b}$ and angle AOB, given that distance $OA = OB = 10$ units.

(b)   Find $\mathbf{c}$, given that OC is perpendicular to l.

(a) $\overrightarrow{OA} = \underline{a} = \begin{bmatrix} 4+3t \\ 0 \\ -2+t \end{bmatrix}$ for some value t.   (1)

Distance $OA = |\mathbf{a}|$.

Since $|\underline{a}| = 10$,

$$\sqrt{(4+3t)^2 + (-2+t)^2} = 10$$

$$16 + 24t + 9t^2 + 4 - 4t + t^2 = 100$$

$$10t^2 + 20t - 80 = 0$$

$$t^2 + 2t - 8 = 0$$

$$(t-2)(t+4) = 0$$

$$\Rightarrow t = 2 \text{ or } t = -4$$

Sample questions and model answers (continued)

This gives the position vectors of **a** and **b**.

Substituting for t in (1) gives

$$\underline{a} = \begin{bmatrix} 10 \\ 0 \\ 0 \end{bmatrix} \quad \text{and} \quad \underline{b} = \begin{bmatrix} -8 \\ 0 \\ -6 \end{bmatrix}$$

Let angle AOB be θ.

$$|\underline{a}| = |\underline{b}| = 10 \quad \text{and} \quad \underline{a} \cdot \underline{b} = \begin{bmatrix} 10 \\ 0 \\ 0 \end{bmatrix} \cdot \begin{bmatrix} -8 \\ 0 \\ -6 \end{bmatrix} = -80$$

$$\underline{a} \cdot \underline{b} = |\underline{a}||\underline{b}| \cos \theta$$

$$\implies \cos \theta = \frac{-80}{100} = -0.8$$

$$\implies \theta = 143.1° \text{ (1 d.p.)}$$

$\begin{bmatrix} 3 \\ 0 \\ 1 \end{bmatrix}$ is the direction vector of l and the scalar product of perpendicular vectors is zero.

(b) $\quad \underline{c} = \begin{bmatrix} 4+3t \\ 0 \\ -2+t \end{bmatrix}$ for some value t. $\hspace{2cm}$ (2)

Since $\underline{c}$ is perpendicular to l,

$$\begin{bmatrix} 4+3t \\ 0 \\ -2+t \end{bmatrix} \cdot \begin{bmatrix} 3 \\ 0 \\ 1 \end{bmatrix} = 0$$

$$\implies 12 + 9t - 2 + t = 0$$

$$\implies t = -1$$

Substituting for t in (2) gives $\underline{c} = \begin{bmatrix} 1 \\ 0 \\ -3 \end{bmatrix}$.

Practice examination questions

Core 4

1 It is given that $f(x) = \dfrac{1}{(1+x)^2} + \sqrt{4+x}$.

Show that if x^3 and higher powers are ignored, $f(x) \approx a + bx + cx^2$ and find the values of a, b and c.

State the values of x for which the expansion is valid.

Core 3

2 (a) Find $\dfrac{dy}{dx}$ when $y = \cos^2 x \sin x$, expressing your answer in terms of $\cos x$.

(b) Using the substitution $u = \cos x$, or otherwise, find $\displaystyle\int \cos^2 x \sin x \, dx$.

Core 3

3 A curve has equation $y = \dfrac{x^2 - 4}{x+1}$.

The equation of the normal to the curve at $(2, 0)$ is $ax + by + c = 0$, where a, b and c are integers.

Find the values of a, b, and c.

Core 3

4 The functions f and g are defined by:

$f{:}x \mapsto 5x - 7 \qquad x \in \mathbb{R}$

$g{:}x \mapsto (x+1)(x-1) \quad x \in \mathbb{R}$.

(a) Find the range of g.

(b) Find an expression for $fg(x)$.

(c) Determine the values of x for which $fg(x) = f(x)$.

Core 4

5 (a) Express $15 \sin x + 8 \cos x$ in the form $R \sin(x + \alpha)$ where $0 < \alpha < 90°$.

(b) State the maximum value of $15 \sin x + 8 \cos x$.

(c) Solve the equation $15 \sin x + 8 \cos x = 12$.
Give all the solutions in the interval $0 < x < 360°$.

Core 4

6 Expressing y in terms of x, solve the differential equation

$$x \frac{dy}{dx} = y + yx,$$

given that $y = 4$ when $x = 2$.

Core 3

7 (a) Use the chain rule to differentiate:

    (i) $y = (4x - 3)^9$

    (ii) $y = \ln(5 - 2x)$

(b) Using your results from part (a):

    (i) Write down an expression for $\displaystyle\int (4x - 3)^8 \, dx$

    (ii) Calculate $\displaystyle\int_3^4 \frac{1}{5 - 2x} \, dx$

Practice examination questions (continued)

Core 3

8   (a)   Use integration by parts to evaluate $\displaystyle\int_0^1 x\,e^{-2x}\,dx$.

(b)   The diagram shows the graph of $y = x\,e^{-x}$.

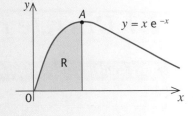

The curve has a maximum point at A.

The shaded region R is the area bounded by the curve, the x-axis and the vertical line through A.

(i)   Show that the coordinates of A are $(1,\,e^{-1})$.
(ii) The region R is rotated completely about the x-axis.
      Calculate the volume of the solid generated.

Core 4

9   In a biology experiment, the growth of a population is being investigated.

A simple model is set up in which the rate of increase of the population at time t is proportional to the number, P, in the population at that time.

It is known that when $t = 0$, $P = 500$ and when $t = 10$, $P = 1000$.

Obtain the relationship between P and t and estimate the size of the population, according to the model, when $t = 20$.

Core 4

10   (a)   Find the equation of the normal to the curve $x^3 + xy + y^2 - 7$ at the point $(1,\,2)$.

(b)   Find the coordinates of the two stationary points on the curve $x = t^2$, $y = t + \dfrac{1}{t}$.

Core 3

11   The region R, shaded in the diagram, is bounded by the curve $y = \sin 2x$ and the x-axis.

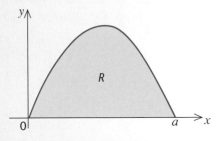

(a)   Find the value of a.

(b)   Show that the area of R is 1 square unit.

Part (c) Core 4

(c)   Find the volume of the solid formed when R is rotated completely about the x-axis.

Core 4

12   (a)   Find a vector equation of the line AB through $A(2,\,2,\,0)$ and $B(2,\,3,\,1)$.

(b)   Show that the line AB and the line $\mathbf{r} = \begin{bmatrix} 0 \\ 2 \\ 1 \end{bmatrix} + \lambda \begin{bmatrix} 1 \\ 0 \\ 1 \end{bmatrix}$, where λ is a scalar parameter, have no common point.

(c)   Find the acute angle between the two lines.

The following topics are covered in this chapter:

- *Discrete random variables*
- *Continuous random variables*
- *Hypothesis testing 1*
- *Hypothesis testing 2*

2.1 Discrete random variables

After studying this section you should be able to:

LEARNING SUMMARY

- *calculate the mean, variance and standard deviation of a discrete random variable and of simple functions of a discrete random variable*
- *understand the Poisson distribution*

Mean, variance and standard deviation

The score obtained when a dice is thrown is an example of a **random variable**. Since only specific values may be obtained, it is a **discrete random variable**.

In the usual notation:
- A capital letter such as X is used to denote the random variable.
- A lower case letter such as x is used to represent a particular value of X.

> This is a useful result.

For a **discrete random variable** X with **probability function** $P(X = x)$, the sum of the probabilities is 1, i.e. $\Sigma P(X = x) = 1$.

> This is usually shown in a table.

The **probability distribution** of X is given by the set of all possible values of x, together with the values of $P(X = x)$.

The **mean of X**, μ, is written $E(X)$ and is also known as the **expected value, or expectation, of X**. If $g(x)$ is a **function of X**, the mean of $g(X)$ is written $E(g(X))$.

> $\sigma = \sqrt{\mathrm{Var}(X)}$

The **standard deviation of X** is σ and the **variance of X** is written $\mathrm{Var}(X)$, so $\mathrm{Var}(X) = \sigma^2$.

KEY POINT

Mean of X:	$\mu = E(X) = \Sigma x P(X = x)$
Mean of $g(X)$:	$E(g(X)) = \Sigma g(x) P(X = x)$
Variance of X:	$\mathrm{Var}(X) = E(X^2) - \mu^2$ where $E(X^2) = \Sigma x^2 P(X = x)$
Alternatively	$\mathrm{Var}(X) = E[(X - \mu)^2] = \Sigma(x - \mu)^2 P(X = x)$

Example

The table shows the probability distribution for a random variable X.

> Notice that
> $\Sigma P(X = x) = 1$.

x	1	2	3
$P(X = x)$	0.2	0.5	0.3

Calculate
(a) $P(X > 1)$ (b) $E(X)$ (c) The standard deviation of X.

(a) $P(X > 1) = 0.5 + 0.3 = 0.8$

(b) $E(X) = \Sigma x P(X = x) = 1 \times 0.2 + 2 \times 0.5 + 3 \times 0.3 = 2.1$

(c) $E(X^2) = \Sigma x^2 P(X = x) = 1^2 \times 0.2 + 2^2 \times 0.5 + 3^2 \times 0.3 = 4.9$
$\mathrm{Var}(X) = E(X^2) - [E(X)]^2 = 4.9 - 2.1^2 = 0.49$
Standard deviation $\sigma = \sqrt{0.49} = 0.7$

For **linear** functions of X:
$E(aX \pm b) = aE(X) \pm b$
$Var(aX \pm b) = a^2Var(X)$

The **mean and variance** of **simple functions of random variables** may be found as in the following examples:

$E(4X + 5) = 4E(X) + 5$ $\qquad$ $Var(4X + 5) = 4^2Var(X) = 16Var(X)$

$E(3X^{-2} - 2) = 3E(X^{-2}) - 2$ $\qquad$ $Var(3X^{-2} - 2) = 3^2Var(X^{-2}) = 9Var(X^{-2})$

The Poisson distribution

λ is the parameter of the distribution.

The **Poisson distribution** is used to model the number of occurrences, X, of an event in a given interval of space or time, when the events occur **independently and at random, at a constant average rate**.

If the mean number of occurrences in the interval is λ, then $X \sim Po(\lambda)$.

Mean = variance

> **KEY POINT**
>
> If $X \sim Po(\lambda)$:
>
> $$P(X = x) = e^{-\lambda}\frac{\lambda^x}{x!} \qquad \text{for } x = 0, 1, 2, 3, \ldots$$
>
> $\mu = E(X) = \lambda$
> $\sigma^2 = Var(X) = \lambda$ and $\sigma = \sqrt{\lambda}$
>
> If $X \sim Po(\lambda_1)$ and $Y \sim Po(\lambda_2)$, then $X + Y \sim Po(\lambda_1 + \lambda_2)$, provided that X and Y are independent.

This is similar to using cumulative Binomial tables in Statistics 1.

For some values of λ you can use **cumulative Poisson probability tables** to find $P(X \leq x)$ and from this you can find other probabilities, for example:

$P(X < 2) = P(X \leq 1)$ $\qquad$ **0 1** 2 3 $\ldots$
$P(X > 4) = 1 - P(X \leq 4)$ $\qquad$ 0 1 2 3 4 **5 6 7** $\ldots$
$P(X \geq 3) = 1 - P(X \leq 2)$ $\qquad$ 0 1 2 **3 4 5** $\ldots$
$P(X = 4) = P(X \leq 4) - P(X \leq 3)$ $\qquad$ 0 1 2 3 **4** 5 $\ldots$

Example
The number of letters received by the Adams family on a weekday may be modelled by a Poisson distribution with mean 3.5. Determine the probability that:
(a) 4 letters are received on Monday
(b) a total of more than 8 letters are received on Thursday and Friday.

$\lambda = 3.5$ is not in the Poisson tables so you have to use the formula.

(a) X is the number of letters received on Monday.

$X \sim Po(3.5)$ so $P(X = 4) = e^{-3.5}\frac{3.5^4}{4!} = 0.18881\ldots = 0.189$ (3 s.f.)

$\lambda = 7$ is in the Poisson tables.

(b) T is the total number of letters received on Thursday and Friday.
$T \sim Po(3.5 + 3.5)$, i.e. $T \sim Po(7)$.
$P(T > 8) = 1 - P(T \leq 8) = 1 - 0.7291 = 0.2709 = 0.271$ (3 s.f.)

Progress check

1 The discrete random variable X has probability distribution as shown:

x:	0	1	2	3
$P(X = x)$	0.3	0.2	k	0.1

Find: (a) k $\quad$ (b) the mean of X $\quad$ (c) the standard deviation of X.

2 A rectangle has sides Y and $(5 + 30Y^{-1})$. Given that $E(Y) = 7$ and $Var(Y) = 20$, find the values of (a) the mean and (b) the variance of the area of the rectangle.

3 The variable X follows a Poisson distribution with standard deviation 2.
(a) Write down the value of λ.
(b) Use cumulative probability tables to find:
(i) $P(X < 5)$ $\quad$ (ii) $P(X \geq 1)$ $\quad$ (iii) $P(X = 6)$.

4   The number of telephone calls to a switchboard may be modelled by a Poisson distribution with mean 3.9 calls per 5-minute interval. Determine, to 3 decimal places, the probability that there will be:
(a) at least one call in a 5-minute interval
(b) exactly 9 calls in a 10-minute interval.

4 (a) 0.980   (b) 0.121
3 (a) 4   (b) (i) 0.6288   (ii) 0.9817   (iii) 0.1042
2 (a) 45   (b) 500
1 (a) 0.4   (b) 1.3   (c) 1.005 (3 d.p.)

2.2 Continuous random variables

- find probabilities and the mean and variance of a continuous variable with probability density function $f(x)$
- find the cumulative function $F(x)$ and the median and quartiles
- find the probability density function from the cumulative function
- understand the continuous rectangular (uniform) distribution

LEARNING SUMMARY

Mean, variance and standard deviation

The continuous random variable X is defined by its **probability density function** $f(x)$, together with the values for which it is valid.

Probabilities are given by areas under the curve.

$$P(c \leqslant X \leqslant d) = \int_c^d f(x)\,dx$$

The total probability is 1.

If $f(x)$ is valid for $a \leqslant x \leqslant b$, then $\int_a^b f(x)\,dx = 1$.

$$P(X = c) = 0$$

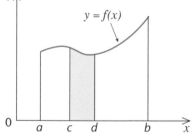

The **expectation** of $g(x)$ is $E(g(x)) = \int_{all\ x} g(x)f(x)\,dx$, for any function $g(x)$.

The functions used most often are X and X^2 as $E(X)$ is the mean and $E(X^2)$ is used to find the variance.

Mean:	$E(X) = \mu = \displaystyle\int_{all\ x} xf(x)\,dx$
Variance:	$Var(X) = \sigma^2 = E(X^2) - \mu^2$, where $E(X^2) = \displaystyle\int_{all\ x} x^2 f(x)\,dx$

KEY POINT

The mean is the **expectation** of X, written $E(X)$. It is often denoted by μ.

If $f(x)$ is given in stages, then the integration is carried out in stages also.

Example

The continuous random variable X has probability density function $f(x)$ given by $f(x) = \frac{1}{4}x$ for $1 \leqslant x \leqslant 3$. Find the mean μ, $E(X^2)$ and the standard deviation σ.

$$\mu = E(X) = \int_{all\ x} xf(x)\,dx = \int_1^3 \frac{1}{4}x^2\,dx = \left[\frac{x^3}{12}\right]_1^3 = 2\tfrac{1}{6}$$

$$E(X^2) = \int_{all\ x} x^2 f(x)\,dx = \int_1^3 \frac{1}{4}x^3\,dx = \left[\frac{x^4}{16}\right]_1^3 = 5$$

$\sigma = \sqrt{variance}$

$$Var(X) = 5 - (2\tfrac{1}{6})^2 = \tfrac{11}{36} \Rightarrow \sigma = \sqrt{\tfrac{11}{36}} = 0.553 \text{ (3 d.p.)}$$

The relationships relating to the **mean and variance of a simple function of a continuous random variable** are the same as those for a discrete random variable. See page 64.

The cumulative distribution function, $F(x)$

The **cumulative distribution function**, $F(x)$, gives the probability that X is less than a particular value, i.e. $P(X < x)$. It is found from the area under the probability curve:

$F(x_1) = P(X < x_1)$.

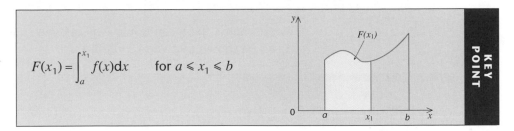

$$F(x_1) = \int_a^{x_1} f(x)\,dx \qquad \text{for } a \leqslant x_1 \leqslant b$$

KEY POINT

Sometimes the median is denoted by q_2.

The **median**, m, is the value 50% of the way through the distribution. It divides the total area in half, so $F(m) = 0.5$.

The **lower quartile** q_1 is the value 25% of the way, so $F(q_1) = 0.25$
The **upper quartile** q_3 is the value 75% of the way, so $F(q_3) = 0.75$

The **interquartile range** = upper quartile – lower quartile.

Example

$f(x) = \frac{1}{4} x$ for $1 \leqslant x \leqslant 3$. Find the median, m, and the interquartile range.

x_1 is used here to denote any value of x between 1 and 3. Sometimes a different letter is used, such as t.

For $1 \leqslant x_1 \leqslant 3$, $F(x_1) = \int_1^x \frac{1}{4} x\,dx = \left[\dfrac{x^2}{8}\right]_1^{x_1} = \frac{1}{8}(x_1^2 - 1)$

$F(m) = 0.5 \Rightarrow \frac{1}{8}(m^2 - 1) = 0.5 \Rightarrow m^2 = 5,\ m = \sqrt{5}\ (= 2.236\ ...)$
$F(q_1) = 0.25 \Rightarrow \frac{1}{8}(q_1^2 - 1) = 0.25 \Rightarrow q_1^2 = 3,\ q_1 = \sqrt{3}\ (= 1.732\ ...)$
$F(q_3) = 0.75 \Rightarrow \frac{1}{8}(q_3^2 - 1) = 0.75 \Rightarrow q_3^2 = 7,\ q_3 = \sqrt{7}\ (= 2.645\ ...)$
Interquartile range $= q_3 - q_1 = \sqrt{7} - \sqrt{3} = 0.91$ (2 d.p.)

Differentiating the cumulative function gives the probability function.

The probability density function can be obtained from the cumulative distribution function:

$$f(x) = \frac{d}{dx} F(x).$$

KEY POINT

Example

Find $f(x)$ where $F(x) = \begin{cases} 0 & x \leqslant 0 \\ \frac{1}{8} x^3 & 0 \leqslant x \leqslant 2 \\ 1 & x \geqslant 2 \end{cases}$

$f(x) = \dfrac{d}{dx}(\tfrac{1}{8} x^3) = \tfrac{3}{8} x^2$ for $0 \leqslant x \leqslant 2$ and $f(x) = 0$ otherwise.

The rectangular distribution

This is also known as the **uniform** distribution.

If $f(x)$ is distributed **uniformly** in the interval $a \leqslant x \leqslant b$, then $f(x) = \dfrac{1}{b-a}$.

Remember that the total area under the line is 1.

By symmetry, the mean is mid-way between a and b, i.e. $E(X) = \frac{1}{2}(a+b)$.
It can be shown that $\mathrm{Var}(X) = \frac{1}{12}(b-a)^2$.

Progress check

1 $f(x) = 1 - 0.5x,\ 0 \leqslant x \leqslant 2$.

(a) Find $E(X)$, $E(X^2)$ and $\mathrm{Var}(X)$.

(b) Find the cumulative function $F(x)$, the median and the interquartile range.

2 $f(x)$ is distributed uniformly in the interval $1 \leqslant x \leqslant 6$. Find:

(a) $P(1.8 < X < 3.2)$ (b) $E(X)$ (c) the standard deviation of X.

3 A continuous random variable X has cumulative distribution function $F(x)$ as shown. Find the probability density function $f(x)$.

$$F(x) = \begin{array}{ll} 0 & x < 2 \\ \frac{1}{20}x^2 + \frac{1}{5}x - \frac{3}{5} & 2 \leqslant x \leqslant 4 \\ 1 & x > 4 \end{array}$$

3 $f(x) = \frac{1}{10}x + \frac{1}{5}$ for $2 \leqslant x \leqslant 4$ and $f(x) = 0$ otherwise

2 (a) 0.28 (b) 3.5 (c) 1.44 (2 d.p.)

1 (a) $\frac{2}{3}, \frac{2}{3}, \frac{5}{9}$ (b) $F(x) = 0,\ x \leqslant 0,\ F(x) = x - 0.25x^2,\ 0 \leqslant x < 2,\ F(x) = 1,\ x \geqslant 2$; median = 0.586 (3 d.p.),
IQR = 0.732 (3 d.p.)

2.3 Hypothesis testing 1

After studying this section you should be able to:

- *understand the language used in hypothesis testing*
- *understand Type I and Type II errors*
- *understand how to obtain critical z-values*
- *perform the z-test for the mean*
- *perform the z-test for the mean of a normal distribution with known variance*
- *perform the z-test for the mean of a distribution using a normal approximation, using large samples only*

LEARNING SUMMARY

The language of hypothesis testing

When you want to know something about a population, you might perform a **hypothesis** or **significance test**.

A **null hypothesis**, H_0 is made about the population.
Then the **alternative hypothesis H_1**, the situation when H_0 is not true, is stated.

Example

When investigating the mean of a normal distribution, you might make the null hypothesis that the mean is 5. This is written $H_0: \mu = 5$.

The alternative hypothesis would be one of the following:

> Use $\mu > 5$ if you suspect an *increase*.
>
> Use $\mu \neq 5$ if you suspect a *change*.

- $H_1: \mu > 5$ (one-tailed, upper tail test)
- $H_1: \mu < 5$ (one-tailed, lower tail test)
- $H_1: \mu \neq 5$ (two-tailed test).

> Use $\mu < 5$ if you suspect a *decrease*.

> The formula for the test statistic depends on the sampling distribution. Examples are shown in the following text.

A **test statistic** is defined and its distribution when H_0 is true is stated.
Its value, based on information from a random sample taken from the population, is found. This is the **test value**.

The hypothesis test involves deciding whether or not the test value could have come from the distribution defined by the null hypothesis.
If the test value is in the main bulk of the distribution (the **acceptance region**) it is *likely* to have come from the distribution. If it is in the 'tail' end (the **rejection** or **critical region**), it is *unlikely* to have come from the distribution.

> Probability theory is used to decide the placing of the boundary between 'likely' and 'unlikely'.

The decision rule (**rejection criterion**) is based on the **significance level** of the test, which is used to fix the boundaries for the rejection region. The boundaries are called **critical values**.

The significance level is a given as a percentage.

> The levels used most often are 10%, 5% and 1%.

For **example**, for a significance level of 5%, the critical values are such that 5% of the distribution is in the critical (rejection) region,

> There are two critical values in a two-tailed test.

i.e. P(X > critical value) = 0.05, for a one-tailed, upper tail test
P(X < critical value) = 0.05, for a one-tailed, lower tail test
P(X > upper critical value) = 0.025, for a two-tailed test
P(X < lower critical value) = 0.025, for a two-tailed test.

> A test statistic is said to be significant if it lies in the critical region. If it lies in the acceptance region, then it is not significant.

Depending on the position of the test value, the **decision** is made.
If the test value lies in the critical region, H_0 is rejected in favour of H_1.
If the test value is in the acceptance region, H_0 is not rejected.

The **conclusion** is then stated in relation to the situation being tested.

Type I and Type II errors

You make a Type I error when you reject a true null hypothesis.

In a hypothesis test, either H_0 is rejected or H_0 is not rejected.
There are occasions when the decision is incorrect and an error is made.

If H_0 is rejected when it is in fact true, a **Type I** error is made.

The probability of making a Type I error is the same as the significance level of the test. For example, if the significance level is 5%, then P(Type I error) = 0.05.

You make a Type II error when you accept a false null hypothesis.

If H_0 is accepted (i.e. not rejected) when it is in fact false, a **Type II** error is made.

To find the probability of making a Type II error, a specific value for H_1 must be given. Then P(Type II error) = P(H_0 is accepted when H_1 is true).

Critical values for z-tests

Some hypothesis tests involving the normal distribution are known as z-tests. In a z-test, the critical values for the rejection region can be found using the standard normal distribution, Z. The most commonly used values are summarised below.

5%
For one-tailed upper tail, $\Phi(z) = 0.95$
For two tailed, upper tail, $\Phi(z) = 0.975$
Use symmetry for lower tails.

10%
For one-tailed upper tail, $\Phi(z) = 0.90$
For two tailed, upper tail $\Phi(z) = 0.95$

1%
For one-tailed upper tail, $\Phi(z) = 0.99$
For two tailed, upper tail $\Phi(z) = 0.995$

Level	Type of test	Critical value	Decision rule
5%	One-tailed (upper)	1.6449	Reject H_0 if $z > 1.6449$
	One-tailed (lower)	−1.6449	Reject H_0 if $z < -1.6449$
	Two-tailed	±1.96	Reject H_0 if $z > 1.96$ or $z < -1.96$
10%	One-tailed (upper)	1.2816	Reject H_0 if $z > 1.2816$
	One-tailed (lower)	−1.2816	Reject H_0 if $z < -1.2816$
	Two-tailed	±1.6449	Reject H_0 if $z > 1.6449$ or $z < -1.6449$
1%	One-tailed (upper)	2.3263	Reject H_0 if $z > 2.3263$
	One-tailed (lower)	−2.3263	Reject H_0 if $z < -2.3263$
	Two-tailed	±2.5758	Reject H_0 if $z > 2.5758$ or $z < -2.5758$

z-test for the mean

This is used to test the mean μ of a distribution when the **variance** σ^2 **is known**.
If the distribution is **normal**, the sample can be **any size**.
If the distribution is **not normal** then **large samples** must be taken so that the **central limit theorem** can be applied to the distribution of the mean (see Revise AS pages 93–94).

When the population is not normal, the central limit theorem is needed.

KEY POINT

Test statistic $Z = \dfrac{\bar{X} - \mu}{\sigma/\sqrt{n}} \sim N(0, 1)$.

Provided n is large ($n > 30$ say), this formula can also be used when the **variance of X is unknown**. In this case the **unbiased estimate** s^2 is used for σ^2, where $s^2 = \dfrac{\Sigma(x - \bar{x})^2}{n - 1}$. If you do know the raw data or summary statistics, use the sample standard deviation, $x\sigma_{n-1}$ on calculator.

Example

A machine fills tins of baked beans such that the mass of a filled tin is normally distributed with mean mass 429 g and the standard deviation is 3 g. The machine breaks down and after being repaired, the mean mass of a random sample of 100 tins is found to be 428.45 g. Is this evidence, at the 5% level, that the mean mass of tins filled by this machine has decreased? Assume that the standard deviation remains unaltered.

Define your variables.

Let X be the mass, in grams, of a filled tin and let the mean after the repair be μ.

State the hypotheses.

$H_0\colon \mu = 429$; $H_1\colon \mu < 429$.

State the distribution of the test statistic assuming the null hypothesis is true.

Consider the sampling distribution of means $\bar{X} \sim \mathrm{N}\left(429, \dfrac{3^2}{100}\right)$.

State the type of test and the rejection criterion.

Perform a one-tailed (lower tail) test at the 5% level and reject H_0 if $z < -1.6449$

Calculate the value of the test statistic.

$\bar{x} = 428.45$, so

$$z = \frac{\bar{x} - \mu}{\sigma/\sqrt{n}} = \frac{428.45 - 429}{3/\sqrt{100}} = -1.833\ldots$$

Decide whether to reject H_0 or not and relate your conclusion to the question.

Since $z < -1.6449$, the test value lies in the critical region and H_0 is rejected.

There is evidence, at the 5% level, that the **mean** mass of a filled tin has decreased.

Alternative method.

Alternatively, a **probability method** can be used.
The rejection rule is to reject H_0 if $P(\bar{X} < 428.45) < 0.05$

$$P(\bar{X} < 428.45) = P\left(Z < \frac{428.45 - 429}{3/\sqrt{100}}\right)$$
$$= P(Z < -1.833\ldots) = 0.03$$

Since $P(\bar{X} < 428.45) < 0.05$, H_0 is rejected and the conclusion is as above.

Progress check

1 $X \sim \mathrm{N}(\mu, 4)$. The mean of a random sample of 10 items from X is 20.9. Test, at the 5% level, the hypotheses $H_0\colon \mu = 20$, $H_1\colon \mu > 20$.

2 The heights of a particular plant have mean μ and variance σ^2 (both unknown). The heights H, in centimetres, of a random sample of 100 plants are summarised as follows: $\Sigma h = 2941$, $\Sigma(h - \bar{h})^2 = 2007.19$

(a) Find unbiased estimates for μ and σ.

(b) Does the sample data support the claim that the mean height is less than 30 cm? Perform a hypothesis test at the 10% level and state your hypotheses.

2 (a) 29.41, 4.503   (b) $H_0\colon \mu = 30$, $H_1\colon \mu < 30$; $z = -1.310$, lies in critical region so reject H_0 and uphold claim.

1 $z = 1.423$, do not reject H_0.

2.4 Hypothesis testing 2

After studying this section you should be able to:

- understand the *t*-distribution
- use the *z*-distribution to obtain confidence intervals for the mean of a normal distribution with known variance or the mean of any distribution based on a large sample
- use the *t*-distribution to obtain confidence intervals for the mean of a normal distribution with unknown variance, based on a small sample
- perform a *t*-test for the mean of a normal population with unknown variance, using a small sample
- perform a χ^2 contingency table test

LEARNING SUMMARY

The t-distribution

If you take a **small** sample from a **normal population** with **unknown mean** μ and **unknown variance**, σ^2, the *t*-distribution is needed to obtain confidence intervals for μ and perform significance tests.

> For large values of v, the *t*-distribution approaches a normal distribution.

The *t*-**distribution** has a similar shape bell shape to the normal distribution. It has one parameter, v, known as the number of **degrees of freedom**.

For a particular value of v, the appropriate *t*-distribution is denoted by $t(v)$. Tables give values of t for which $P(T < t) = p$, for various values of p, usually $p = 0.75$, 0.90, 0.95, 0.975, 0.99.

Estimation – confidence intervals

> See Revise AS page 94.

For samples, size n, taken from a **normal distribution** $X \sim N(\mu, \sigma^2)$ with σ **known**:

> This also holds when large samples are taken from any distribution. In this case, if σ is unknown, use s instead.

The 95% **confidence limits** for μ are $\bar{x} \pm 1.96 \dfrac{\sigma}{\sqrt{n}}$, where $(-1.96, 1.96)$ encloses the central 95% of the $Z \sim N(0, 1)$ distribution. The 95% **confidence interval** for μ is

$$(\bar{x} - 1.96 \frac{\sigma}{\sqrt{n}}, \bar{x} + 1.96 \frac{\sigma}{\sqrt{n}}).$$

KEY POINT

For **small** samples, size n, taken from a **normal** distribution $X \sim N(\mu, \sigma^2)$ with σ **unknown**:

The 95% **confidence limits** for μ are $\bar{x} \pm t \dfrac{s}{\sqrt{n}}$,

where $(-t, t)$ encloses the central 95% of the $t(n-1)$ distribution.

The 95% **confidence interval** for μ is $\left(\bar{x} - t\dfrac{s}{\sqrt{n}}, \bar{x} + t\dfrac{s}{\sqrt{n}}\right)$.

KEY POINT

Example

> When the sample is small, the variable must follow a normal distribution.

A random sample of 8 observations from $X \sim N(\mu, \sigma^2)$ gave
$\Sigma x = 36.4$, $\Sigma(x - \bar{x})^2 = 22.46$

(a) Find an unbiased estimate of σ.
(b) Find a 95% confidence interval for μ.

(a) $s^2 = \dfrac{\Sigma(x - \bar{x})^2}{n - 1} = \dfrac{22.46}{7} = 3.208 \ldots$

so $s = \sqrt{3.208 \ldots} = 1.791 \ldots$

(b) Since $v = 7$, use the $t(7)$ distribution.

Since the confidence interval is symmetric (two-tailed), find $v = 7$, $p = 0.975$ (2.5% in the upper tail). This gives critical value 2.365.

> You need to find t such that $P(-t < T < t) = 0.95$, i.e. $P(T < t) = 0.975$.

The 95% confidence limits for μ are $\bar{x} \pm t\dfrac{s}{\sqrt{n}} = 4.55 \pm 2.365\left(\dfrac{1.791 \ldots}{\sqrt{8}}\right)$

$= 4.55 \pm 1.497 \ldots$

The 95% confidence interval $= (4.55 - 1.497 \ldots, 4.55 + 1.497 \ldots)$

$= (3.05, 6.05)$ (2 d.p.)

t-tests

t-test for the mean

When testing the mean from a normal population with unknown variance, when the sample size is small, the test statistic T follows a $t(n - 1)$ distribution.

> The distribution of X must be normal.

> s is the unbiased estimate of σ.

KEY POINT

Test statistic $T = \dfrac{\bar{X} - \mu}{s/\sqrt{n}} \sim t(n - 1)$.

Example

Using the data given in the preceding example:
test at the 5% level the claim that μ is greater than 4.

> The null hypothesis is $\mu = 4$, even though you are asked to test the claim that $\mu > 4$.

H_0: $\mu = 4$, H_1: $\mu > 4$.

Using t-tables for $t(7)$,

> You want t such that $P(T < t) = 0.95$

the critical value for $v = 7$, $p = 0.95$ (one-tailed test, 5% in the upper tail) is 1.895 so reject H_0 if $t > 1.895$.

$$\bar{x} = \frac{\Sigma x}{n} = \frac{36.4}{8} = 4.55,$$

$$\Rightarrow t = \frac{\bar{x} - \mu}{s/\sqrt{n}} = \frac{4.55 - 4}{1.791/\sqrt{8}} = 0.868 \ldots$$

Since $t < 1.895$, the test value does not lie in the critical region. There is no evidence to support the claim that μ is greater than 4.

If you are given raw data you can obtain s using your calculator by finding $x\sigma_{n-1}$ in SD mode.

χ^2 contingency table test

A χ^2 **contingency table test** can be used to test whether two factors are independent or whether there is an association between them.

This is a 3 by 4 table

Data are in the form of a **contingency table**, an array displaying data relating to the two factors. The size of the table is described by the number of rows and the number of columns. An h by k table has h rows and k columns.

The **observed frequencies**, O, are put into the table.

Here are the stages of the test.

- Make the **null** and **alternative hypotheses**:
  The null hypothesis, H_0, is that the two factors are independent.
  The alternative hypothesis, H_1, is that there is an association between them.

Expected frequencies should be greater than 5, so it may be necessary to combine adjacent rows or columns. This must be done sensibly, with regard to the categories.

- Calculate row totals, column totals and the grand totals for the observed data and use these to calculate the **expected frequencies**, E, if the factors are independent.
  The expected frequency for a particular cell is calculated using

$$E = \frac{(\text{row total}) \times (\text{column total})}{\text{grand total}}$$

  Make a table showing the expected frequencies.

- Work out v, the number of **degrees of freedom**. This depends on the size of the contingency table. For an h by k table, $v = (h-1)(k-1)$.

- Decide on the significance level of the test and look up the critical value in χ^2 tables.

- Calculate the **test statistic** $\chi^2 = \sum \dfrac{(O-E)^2}{E}$.

  For a 2 by 2 contingency table, where $v = 1$, use Yates' continuity correction.
  In this case, $\chi^2 = \sum \dfrac{(|O-E|-0.5)^2}{E}$.

- Make your conclusion.
  If $\chi^2 >$ critical value, reject H_0 and conclude that the factors are not independent; there is an association between them.
  If $\chi^2 <$ critical value, do not reject H_0 and conclude that the factors are independent.

Example

Test, at the 10% significance level, whether size and colour are independent, for the data given in the following table.

This is a 3 by 2 contingency table.

		Colour		
		Pink	White	Totals
Size	Small	12	23	35
	Medium	16	20	36
	Large	22	17	39
	Totals	50	60	110

H_0: Size and colour are independent;
H_1: Size and colour are not independent

Expected frequencies (to 1 decimal place).

Small pink: $\quad E = \dfrac{(\text{row total}) \times (\text{column total})}{\text{grand total}} = \dfrac{35 \times 50}{110} = 15.9$

Medium pink: $\quad E = \dfrac{36 \times 50}{110} = 16.4$

Only two expected frequencies need to be calculated using the formula. All the rest can then be found by ensuring that row or column totals agree. This confirms that $v = 2$.

Large pink: $\quad E = 50 - (15.9 + 16.4) = 17.7$ (column totals agree).
Small white: $\quad E = 35 - 15.9 = 19.1$ (row totals agree).
Medium white: $\quad E = 36 - 16.4 = 19.6$ (row totals agree).
Large white: $\quad E = 39 - 17.7 = 21.3$ (row totals agree).

$v = (3-1)(2-1) = 2 \times 1 = 2$. Consider the $\chi^2(2)$ distribution. At the 10% level, the critical value (from tables) is 4.605, so H_0 is rejected if $\chi^2 > 4.605$

O	E	$\dfrac{(O-E)^2}{E}$
12	15.9	0.956 …
16	16.4	0.0097 …
22	17.7	1.044 …
23	19.1	0.796 …
20	19.6	0.00816 …
17	21.3	0.868 …
$\Sigma O = 160$	$\Sigma E = 160$	$\chi^2 = 3.6835 …$

$$\chi^2 = \sum \frac{(O-E)^2}{E} = 3.68$$

Since $\chi^2 < 4.605$, H_0 is not rejected.
Size and colour are independent.

Progress check

1 It is known that the times, in seconds, taken to perform a particular task are normally distributed with mean μ and variance σ^2.
The times, in seconds, taken by 10 volunteers are as follows:
32.9, 27.6, 33.0, 35.7, 39.2, 26.8, 34.3, 31.1, 38.9, 28.7

(a) Find an unbiased estimate of σ.
(b) Find a 99% confidence interval for μ.
(c) Find the width of the 95% confidence interval.
(d) Test at the 5% level the null hypothesis that $\mu = 35.5$, against the alternative hypothesis that $\mu < 35.5$.

2 In a random sample of 100 students at a driving school, 28 of the 40 men passed their test first time and 34 of the 60 women passed first time.
(a) Form a 2 × 2 contingency table to show the data.
(b) Test, at the 5% level of significance, the claim that there is a relationship between the ability to pass the driving test first time and whether the candidate is male or female.

3 Use a χ^2 test, at the 5% level of significance, to investigate the claim that there is an association between the bus route and the punctuality of the buses.

		Punctuality			
		On time	Early	Late	Totals
Bus route	Route 1	90	35	85	210
	Route 2	30	30	60	120
	Route 3	30	15	25	70
	Totals	150	80	170	400

1 (a) 4.37 (b) (28.3, 37.3) (c) 6.25
   (d) $t = -1.941$, reject H_0 since $t < -1.833$; mean time is less than 35.5 seconds.

2 (a)

	Pass	Fail
Male	28	12
Female	34	26

   (b) $v = 1$, E 24.8, 15.2, 37.2, 22.8; Use Yates' correction $\chi^2 = 1.29 < 3.841$; no evidence of relationship.

3 $v = 4$, $E = 78.75, 42, 89.25, 45, 24, 51, 26.25, 14, 29.75$, $X^2 = 12.43 > 9.488$, there is an association.

Sample questions and model answers

1

The continuous random variable X has probability density function $f(x)$ given by

$$f(x) = \begin{cases} kx(2-x) & 0 \leqslant x \leqslant 2 \\ 0 & \text{otherwise} \end{cases}$$

where k is a constant.

(a) Show that $k = \frac{3}{4}$ and find the exact value of $P(X > 1\frac{1}{2})$.

(b) Find $E(X)$ and $E(X^2)$.

(c) Show that the standard deviation, σ, is 0.447 correct to three decimal places.

The total area under the curve is 1.

(a)
$$\int_0^2 f(x)dx = 1$$

As k is a constant, it can be taken outside the integration.

$$\Rightarrow k\int_0^2 (2x - x^2)dx = 1$$

$$\Rightarrow k\left[x^2 - \frac{x^3}{3}\right]_0^2 = 1$$

$$\Rightarrow k(4 - \tfrac{8}{3} - 0) = 1$$

$$\Rightarrow k = \tfrac{3}{4}$$

$$f(x) = \tfrac{3}{4}(2x - x^2)$$

$$P(X > 1\tfrac{1}{2}) = \int_{1\frac{1}{2}}^2 f(x)dx$$

$$= \tfrac{3}{4}\int_{1\frac{1}{2}}^2 (2x - x^2)dx$$

Substitute the limits carefully, working in fractions to obtain the exact answer.

$$= \tfrac{3}{4}\left[x^2 - \frac{x^3}{3}\right]_{1\frac{1}{2}}^2$$

$$= \tfrac{3}{4}(4 - \tfrac{8}{3} - (\tfrac{9}{4} - \tfrac{9}{8}))$$

$$= \tfrac{5}{32}$$

If you do not spot that $f(x)$ is symmetrical about $x = 1$, find $E(X)$ by integration.

(b) By symmetry, $E(X) = 1$

$$E(X^2) = \int_0^2 x^2 f(x)dx$$

$$= \tfrac{3}{4}\int_0^2 (2x^3 - x^4)dx$$

$$= \tfrac{3}{4}\left[\frac{2x^4}{4} - \frac{x^5}{5}\right]_0^2$$

$$= \tfrac{3}{4}(8 - \tfrac{32}{5})$$

$$= 1.2$$

$[E(X)]^2$ is often written $E^2(X)$.

(c) $\text{Var}(X) = E(X^2) - [E(X)]^2$

$$= 1.2 - 1$$

$$= 0.2$$

$$\sigma = \sqrt{0.2} = 0.447 \ (3 \text{ d.p.})$$

Sample questions and model answers *(continued)*

2

Andy's transport company attends A breakdowns of its vehicles in a week where A is modelled by a Poisson distribution with parameter 0.9.

Barry's transport company attends B breakdowns of its vehicles in a week where B is modelled by a Poisson distribution with parameter 1.2.

(a) Determine the probability that Andy's company attends more than 1 but at most 4 breakdowns in a week.

(b) Determine the probability that the two companies attend a total of 6 breakdowns in a **fortnight**.

(c) The number of breakdowns per week dealt with by Cathy's transport company over a period of twelve weeks are recorded as:

> 1, 3, 4, 0, 0, 2, 5, 2, 1, 0, 1, 0

(i) Calculate the mean and variance of these data.

(ii) State, giving a reason based on your answer to (c) (i) whether or not a Poisson distribution provides a suitable model for these data.

(a) $A \sim \text{Po}(0.9)$. Using tables:

$$P(1 < A \leqslant 4) = P(A \leqslant 4) - P(A \leqslant 1)$$
$$= 0.9977 - 0.7725$$
$$= 0.2252$$

(b) T is the total number of breakdowns in a fortnight.

$T \sim \text{Po}(0.9 + 0.9 + 1.2 + 1.2)$, i.e. $T \sim \text{Po}(4.2)$

$$P(T = 6) = e^{-4.2} \frac{4.2^6}{6!}$$
$$= 0.114 \text{ (3 s.f.)}$$

(c) (i) Using the calculator, $\bar{x} = 1.58$ (3 s.f.), $s^2 = 1.6764...^2 = 2.81$ (3 s.f.)

(ii) The mean and variance are not similar, so the Poisson distribution is unlikely to be an appropriate model.

Sample questions and model answers (continued)

3

The lengths of a population of snakes are normally distributed with standard deviation 8 cm and unknown mean μ cm.

A random sample of 10 snakes is taken and their mean length calculated.

When a significance test, at the 10% level, is performed, the hypothesis that μ is 37.5 is rejected in favour of the hypothesis that it is greater than 37.5 cm.

(a) What can be said about the value of the sample mean?

(b) Explain briefly what is meant, in the context of this question, by a Type I error, and state the probability of making a Type I error.

(a) If X is the length, in centimetres, then $X \sim N(\mu, 8^2)$.

H_0: $\mu = 37.5$

H_1: $\mu > 37.5$

For the significance test, the sampling distribution of means is considered.

If $\mu = 37.5$, then $\bar{X} \sim N\left(37.5, \dfrac{8^2}{10}\right)$.

From standard normal tables, $\Phi(z) = 0.9 \Rightarrow z = 1.2816$

The critical value for a one-tailed (upper tail) at the 10% level is 1.2816.

H_0 is rejected if z is in the upper tail 10% of the distribution.

Since H_0 is rejected, $z > 1.2816$,

Test statistic is $Z = \dfrac{\bar{X} - \mu}{\sigma/\sqrt{n}}$.

i.e. $\dfrac{\bar{x} - 37.5}{8/\sqrt{10}} > 1.2816$

$\bar{x} > 40.74$

Z: 0   1.2816

$\overline{X}$ 37.5 40.74

10%

(b) A Type I error is made when the null hypothesis, $\mu = 37.5$, is rejected, when the mean population length is in fact 37.5

$P(\text{Type I error}) = 10\%$

The probability of making a Type I error is the same as the level of significance of the test.

Practice examination questions

1 It is given that $X \sim \mathrm{N}(\mu, 16)$.
The null hypothesis $\mu = 20$ is to be tested against the alternative hypothesis $\mu \neq 20$.

The mean of a random sample of 10 observations from X is 17.2.

Test at the 5% level whether this provides evidence that μ is not 20.

2 The continuous random variable X is distributed uniformly in the interval $2 \leqslant x \leqslant 12$.

(a) Find the probability density function, $f(x)$.

(b) Calculate the probability that X lies within one standard deviation of the mean, giving your answer correct to 2 decimal places.

3 The random variable X has a normal distribution with mean μ and variance σ^2.
In order to set up a confidence interval for μ, a random sample of 9 observations of X is taken.

The results are summarised as follows:

$\Sigma x = 255.6 \qquad \Sigma(x - \bar{x})^2 = 113.22$

(a) If $\sigma = 4$, calculate a 90% confidence interval for μ.

(b) If σ is unknown,

(i) calculate an unbiased estimate for σ,

(ii) calculate a 95% confidence interval for μ.

4 The continuous random variable X has probability function $f(x)$ where $f(x) = k(x - x^3)$, $0 \leqslant x \leqslant 1$ and $f(x) = 0$ otherwise.

(a) Show that $k = 4$.

(b) Find $E(X)$.

(c) Find $P(X < 0.5)$.

(d) Find $Var(X)$.

5 The continuous random variable X has cumulative distribution function given by

$$F(x) = \begin{cases} 0 & x \leqslant 0 \\ 2cx & 0 \leqslant x \leqslant 1 \\ c(4x - x^2 - 1) & 1 \leqslant x \leqslant 2 \\ 1 & x \geqslant 2 \end{cases}$$

(a) Show that $c = \frac{1}{3}$.

(b) Show that $P(X > 1\frac{1}{2}) = \frac{1}{12}$.

(c) Find the probability function $f(x)$.

(d) Calculate $E(X)$.

6 A psychologist is investigating whether or not there is any association between whether a person is left- or right-handed and the ability to complete a task involving manual dexterity in a given time. A sample of 200 people gave the following results:

	Completed task	Did not complete task
Right-handed	83	67
Left-handed	37	13

Stating your hypotheses clearly, test, at the 5% level, whether or not there is any evidence of an association between right- or left-handedness and the ability to complete the task in the given time.

7 A village newsagent finds that the demand for the local newspaper on a Monday follows a Poisson distribution with mean 15.

(a) Find the probability that there are fewer than 16 requests for the newspaper on a Monday.

(b) How many copies of the newspaper should there be in stock so that the demand is not satisfied on fewer than 5% of Mondays?

Chapter 3

Mechanics 2

The following topics are covered in this chapter:

- *Linear motion with variable acceleration*
- *Variable acceleration using vectors*
- *Equilibrium of a rigid body*
- *Centre of mass*

- *Uniform circular motion*
- *Further circular motion*
- *Work, energy and power*
- *Elastic springs and strings*

3.1 Linear motion with variable acceleration

After studying this section you should be able to:

- *set up and solve a differential equation representing acceleration in a straight line, where the acceleration is given as a function of displacement or time*
- *solve problems which can be modelled as the linear motion of a particle under the action of a variable force*

Kinematics

Kinematics is the branch of mathematics that deals with the motion of a particle in terms of displacement, velocity and acceleration, without considering the forces that may be required to cause the motion.

For motion in a straight line, the distinction between **distance** and **displacement** is only that displacement may be positive or negative to indicate a sense of direction, whereas distance is always taken to be positive.

The same distinction applies to **speed** and **velocity**. Velocity in a straight line may be positive or negative, depending on the direction of movement, but speed is always taken to be positive.

In the standard notation:

- displacement is represented by x

- velocity is represented by $v = \dot{x} = \dfrac{\mathrm{d}x}{\mathrm{d}t}$.

You may need to set up and solve equations of the form $\dfrac{\mathrm{d}x}{\mathrm{d}t} = f(x)$ or $\dfrac{\mathrm{d}x}{\mathrm{d}t} = f(t)$.

- **Acceleration** is represented by $\ddot{x} = \dfrac{\mathrm{d}v}{\mathrm{d}t} = v\dfrac{\mathrm{d}v}{\mathrm{d}x}$.

> This is an important result and you may need to establish it in the exam.

You should recognise that by the chain rule for differentiation: $\dfrac{\mathrm{d}v}{\mathrm{d}t} = \dfrac{\mathrm{d}v}{\mathrm{d}x} \times \dfrac{\mathrm{d}x}{\mathrm{d}t} = v\dfrac{\mathrm{d}v}{\mathrm{d}x}$.

You may need to set up and solve equations of the form $\dfrac{\mathrm{d}v}{\mathrm{d}t} = f(t)$ or $v\dfrac{\mathrm{d}v}{\mathrm{d}x} = f(x)$.

This requires a differential equation because the acceleration is not constant.

Example

A particle P moves in the direction of the positive x-axis with acceleration $3\sqrt{x}$ ms^{-2} directed away from O, where OP $= x$ metres. When $x = 1$ the speed of P is 2 ms^{-1}. Find its speed when $x = 4$.

See **separation of variables** page 46.

The acceleration is $v\dfrac{dv}{dx} = 3x^{\frac{1}{2}} \implies \displaystyle\int v \, dv = \int 3x^{\frac{1}{2}} \, dx$

This gives: $\qquad\qquad \dfrac{v^2}{2} = 2x^{\frac{3}{2}} + c$

When $x = 1$, $v = 2$ so $2 = 2 + c \implies c = 0$.

so: $\qquad\qquad\qquad v^2 = 4x^{\frac{3}{2}}$

When $x = 4$, $v^2 = 32 \implies v = 4\sqrt{2}$

The particle has speed $4\sqrt{2}$ ms^{-1} when $x = 4$.

Dynamics

Dynamics is concerned with the motion of a particle in response to forces that act upon it. The first step is to identify all of the forces and then use $F = ma$ to set up a differential equation.

Example

A particle P of mass 0.2 kg is attached to one end of a light elastic string. The other end of the string is attached to a fixed point on a smooth horizontal table. P is released from rest and moves along the surface of the table towards a point O.

The tension in the string has magnitude $25x$ N where x metres is the displacement of P from O. Given that $x = 2$ at the point of release, find the speed of P when $x = 1$.

Draw a diagram and label the forces.

In the diagram, x increases from left to right so take this as the positive direction.

Using $F = ma$: $\qquad -25x = 0.2v\dfrac{dv}{dx}$

so: $\qquad\qquad \displaystyle\int -125x \, dx = \int v \, dv$

giving: $\qquad\qquad \dfrac{v^2}{2} = -125\dfrac{x^2}{2} + c$

Simplifying the equation can make it easier to substitute particular values.

This may be written as $v^2 = A - 125x^2$

When $x = 2$, $v = 0$ so $\quad 0 = A - 125 \times 4 \implies A = 500$

When $x = 1$, $\qquad\qquad v^2 = 500 - 125 = 375$

giving: $\qquad\qquad\qquad v = \sqrt{375}$

The speed of P when $x = 1$ is 19.4 ms^{-1} to 3 s.f.

Progress check

1 A particle P of mass 0.4 kg moves in the direction of the positive x-axis with acceleration $\dfrac{5}{x}$ ms^{-2} away from O, where OP $= x$ metres. When $x = 1$ the speed of P is 3 ms^{-1}. Find its speed when $x = 5$.

2 A particle P of mass 0.5 kg moves in a straight line, away from a point O, under the action of a force of magnitude $\dfrac{x^2}{12}$ N directed towards O, where OP $= x$ metres. When $x = 3$ the speed of P is 15 ms^{-1}. Find x when P is brought to rest.

1 5.01 ms^{-1} to 3 s.f.
2 12.7 m to 3 s.f.

3.2 Variable acceleration using vectors

After studying this section you should be able to:

- *differentiate and integrate a vector with respect to time*
- *use differentiation and integration of vectors to solve problems involving variable acceleration*

LEARNING SUMMARY

Vector and scalar quantities

Key points from AS

- **Scalars and vectors**
  Revise AS page 59

KEY POINT

A **scalar** quantity has **magnitude** i.e. size, but not direction. Numbers are scalars and some other important examples are distance, speed, mass and time.

A **vector** quantity has both magnitude and direction. For example, distance in a specified direction is called displacement. Some other important examples are velocity, acceleration, force and momentum.

The magnitude of a vector $\mathbf{r} = a\mathbf{i} + b\mathbf{j}$ is given by $r = \sqrt{a^2 + b^2}$

Differentiation of vectors

A vector of the form $\mathbf{r} = f(t)\mathbf{i} + g(t)\mathbf{j}$ may be differentiated with respect to time to give $\dot{\mathbf{r}} = f'(t)\mathbf{i} + g'(t)\mathbf{j}$

For example, if $\mathbf{r} = 3t^2\mathbf{i} + \sqrt{t}\mathbf{j}$ then $\dot{\mathbf{r}} = 6t\mathbf{i} + \dfrac{1}{2\sqrt{t}}\mathbf{j}$.

> Differentiate the coefficients of **i** and **j** with respect to *t*.

If $\mathbf{r}$ represents the position vector of some moving point at time t then $\dot{\mathbf{r}}$ represents the corresponding velocity at time t.

$\dot{\mathbf{r}}$ may be differentiated to give $\ddot{\mathbf{r}}$ which represents the acceleration of the moving point at time t.

Example

A particle P has position vector $\mathbf{r}$ metres relative to a fixed point O, at time t seconds, where $\mathbf{r} = (2t^2 + 5t)\mathbf{i} + t^3\mathbf{j}$ and $t \geqslant 0$.

Find:

(a) the velocity of P at time t seconds

(b) the speed of P when $t = 3$

(c) the time when P is moving parallel to the vector $\mathbf{i} + 3\mathbf{j}$.

(a) The velocity of P in metres per second at time t is given by $\dot{\mathbf{r}} = (4t + 5)\mathbf{i} + 3t^2\mathbf{j}$.

(b) When $t = 3$, $\dot{\mathbf{r}} = 17\mathbf{i} + 27\mathbf{j}$.

> The speed of P is the magnitude of its velocity.

The speed of P when $t = 3$ is $\sqrt{17^2 + 27^2}$ ms^{-1} = 31.9 ms^{-1} to 3 s.f.

(c) When P is moving parallel to $\mathbf{i} + 3\mathbf{j}$ the component of velocity in the $\mathbf{j}$ direction must be 3 times the component in the $\mathbf{i}$ direction.

This gives:
$$3t^2 = 3(4t + 5)$$
so: $$t^2 = 4t + 5$$
$$t^2 - 4t - 5 = 0$$
$$(t - 5)(t + 1) = 0$$
$$t = 5 \text{ or } t = -1$$

However, $t \geqslant 0$ so the only solution is $t = 5$.

Integration of vectors

In general, $\int (f(t)\mathbf{i} + g(t)\mathbf{j})dt = \left(\int f(t)dt\right)\mathbf{i} + \left(\int g(t)dt\right)\mathbf{j}$

> Don't be put off by the complicated looking formula, just look at what it means.

So, to integrate a vector with respect to t, integrate the coefficients of $\mathbf{i}$ and $\mathbf{j}$ with respect to t. **Note that a constant of integration is needed and that this is itself a vector.**

Example

Find $\mathbf{v} = \int \mathbf{a}\, dt$ given that $\mathbf{a} = 2t\mathbf{i} - 6t^2\mathbf{j}$ and that $\mathbf{v} = 3\mathbf{i} + \mathbf{j}$ when $t = 1$.

$\mathbf{v} = \int (2t\mathbf{i} - 6t^2\mathbf{j})dt = t^2\mathbf{i} - 2t^3\mathbf{j} + \mathbf{c}$

When $t = 1$: $\mathbf{v} = \mathbf{i} - 2\mathbf{j} + \mathbf{c} = 3\mathbf{i} + \mathbf{j}$

so: $\mathbf{c} = 2\mathbf{i} + 3\mathbf{j}$

This gives: $\mathbf{v} = t^2\mathbf{i} - 2t^3\mathbf{j} + 2\mathbf{i} + 3\mathbf{j}$
or $\mathbf{v} = (t^2 + 2)\mathbf{i} + (3 - 2t^3)\mathbf{j}$

In the example, if $\mathbf{a}$ represents acceleration then $\mathbf{v}$ represents velocity.

> **KEY POINT**
>
> Vectors for position $\mathbf{r}$, velocity $\mathbf{v}$ and acceleration $\mathbf{a}$ are linked through differentiation and integration.
>
> $\mathbf{v} = \dot{\mathbf{r}}$ $\qquad \mathbf{v} = \int \mathbf{a}\, dt$ $\qquad$ differentiate
>
> $\qquad\qquad\qquad\qquad\qquad\qquad\qquad \mathbf{r} \quad \mathbf{v} \quad \mathbf{a}$
>
> $\mathbf{a} = \dot{\mathbf{v}} = \ddot{\mathbf{r}}$ $\qquad \mathbf{r} = \int \mathbf{v}\, dt$ $\qquad$ integrate

Example

> 'Initially' means when $t = 0$.

A particle P, initially at rest, has acceleration $\mathbf{a} = 2t\mathbf{i} + 3\sqrt{t}\mathbf{j}$ ms^{-2} at time t seconds.

(a) Find the velocity of P at time t seconds.
(b) Show that P is moving parallel to $\mathbf{i} + \mathbf{j}$ when $t = 4$.

> Note that the formulae for constant acceleration cannot be used here.

> $3 \times t^{\frac{3}{2}} \times \frac{2}{3} = 2t^{\frac{3}{2}}$.

(a) $\mathbf{v} = \int \mathbf{a}\, dt = \int (2t\mathbf{i} + 3\sqrt{t}\mathbf{j})dt$

$\qquad = t^2\mathbf{i} + 2t^{\frac{3}{2}}\mathbf{j} + \mathbf{c}$

When $t = 0$, $\mathbf{v} = 0 \Rightarrow \mathbf{c} = 0$ so, $\mathbf{v} = t^2\mathbf{i} + 2t^{\frac{3}{2}}\mathbf{j}$

(b) When $t = 4$, $\mathbf{v} = 16\mathbf{i} + 16\mathbf{j}$
$\qquad = 16(\mathbf{i} + \mathbf{j})$ so P is moving parallel to $\mathbf{i} + \mathbf{j}$.

Progress check

1 An object has position vector $\mathbf{r} = (2t^3\mathbf{i} - 5t^2\mathbf{j})$ m at time t seconds. Find the acceleration of the object when $t = 3$.

2 A particle P, initially at the point with position vector $(-3\mathbf{i} + 5\mathbf{j})$ m, has velocity $\mathbf{v} = \sqrt{t}\mathbf{i} + 3\mathbf{j}$ at time t seconds. Find the position vector of P when $t = 4$.

2 $\frac{16}{3}\mathbf{i} + 17\mathbf{j}$ 1 $\ddot{\mathbf{r}} = 36\mathbf{i} - 10\mathbf{j}$

3.3 Equilibrium of a rigid body

After studying this section you should be able to:

- *understand the conditions for equilibrium of a rigid body under the action of coplanar forces*
- *solve problems involving coplanar forces*

Equilibrium

Forces that act in the same plane are coplanar.

You may need to show that a rigid body is in **equilibrium** under a system of **coplanar** forces *or* to *use* the fact that a rigid body is in equilibrium to find something about the forces acting on it.

To show that a rigid body is in equilibrium under the action of coplanar forces you need to establish that:

- the vector sum of the forces is zero
- the sum of the moments about some point is zero.

You only need to establish this for *one* point and you can choose any point that is convenient.

If a body is in equilibrium under some coplanar forces then you know that:

- the vector sum of the forces will be zero
- the sum of the moments about any point you choose will be zero.

> **KEY POINT**
>
> The conditions for equilibrium can be used to produce equations. The equations can then be solved to find the unknown values in a problem.

Example

A uniform plank AB of length 5 m and mass 20 kg rests horizontally on two supports, one at each end. A mass of 10 kg is positioned on the plank 1 m from B. Find the reaction at each of the supports. Take $g = 9.8 \text{ ms}^{-2}$.

The first step is to draw a clearly labelled diagram.

The plank is uniform so its weight acts at its mid-point.

The vector sum of the forces is zero so the total upward force must equal the total downward force.

Vertically:
$$\mathbf{R}_A + \mathbf{R}_B = 20g + 10g$$
$$= 30g$$

M(A) gives:
$$5\mathbf{R}_B = 20g \times 2.5 + 10g \times 4$$
$$= 90g$$
$$\mathbf{R}_B = 18g$$
$$= 176.4 \text{ N (reaction at B)}$$

Substitute for $\mathbf{R}_B$:
$$\mathbf{R}_A + 18g = 30g$$
$$\mathbf{R}_A = 12g$$
$$= 117.6 \text{ N (reaction at A)}.$$

Leaning ladders

The situation where a ladder leans against a wall, possibly supporting a load at some point along its length, provides a rich source of questions on equilibrium. It requires forces and moments to be considered and includes the theory of friction.

Example
A uniform ladder of mass m kg rests against a smooth vertical wall with its lower end on rough horizontal ground. A man of mass $4m$ kg has climbed three-quarters of the way up the ladder but cannot go any higher or it will slip. The ladder makes an angle of $60°$ with the horizontal.
Find the coefficient of friction between the ladder and the ground.

Start by drawing a clearly labelled diagram.

The wall is smooth so there is no vertical force at the top of the ladder.

The length of the ladder has been written as $4l$ to simplify taking moments.

Friction acts at the bottom of the ladder to oppose the tendency for it to slip.

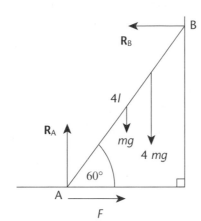

In this example, three equations may be formed by resolving the forces in two directions and taking moments.

The equations provide enough information to solve the problem.

Vertically:
$$\mathbf{R}_A = mg + 4mg \qquad (1)$$
$$= 5mg$$

Horizontally:
$$F = \mathbf{R}_B \qquad (2)$$

M(A):
$$\mathbf{R}_B \times 4l \sin 60° = mg \times 2l \cos 60° + 4mg \times 3l \cos 60°$$

giving:
$$\mathbf{R}_B \times 2\sqrt{3}l = mgl + 6mgl$$

so:
$$\mathbf{R}_B = \frac{7mg}{2\sqrt{3}} \qquad (3)$$

From (2) and (3)
$$F = \frac{7mg}{2\sqrt{3}}$$

It's a good idea to rationalise the denominator by multiplying the top and bottom of the fraction by $\sqrt{3}$.

Since the ladder is on the point of slipping, friction is limiting and the coefficient of friction is given by:

$$\mu = \frac{F}{\mathbf{R}_A} = \frac{7}{10\sqrt{3}} = \frac{7\sqrt{3}}{30}$$

Progress check

1. A uniform plank AB of length 3 m and mass 25 kg rests horizontally on two supports, one at each end. A mass of 15 kg is positioned on the plank 1 m from B. Find the reaction at each of the supports. Take $g = 9.8$ ms^{-2}.

2. A uniform ladder of length 5 m rests against a smooth wall with its base on rough horizontal ground. The coefficient of friction between the ladder and the ground is $\frac{2}{3}$ and the ladder is on the point of slipping. Given that the angle between the ladder and the ground is x, find $\tan x$.

2 $\frac{3}{4}$

1 220.5 N (A), 171.5 N (B)

3.4 Centre of mass

After studying this section you should be able to:

- *find the centre of mass of a system of particles in one and two dimensions*
- *find the centre of mass of a lamina by considering an equivalent system of particles*
- *use centre of mass to determine conditions for the equilibrium of a lamina*

LEARNING SUMMARY

Centre of mass of a system of particles

In the diagram, m_1 and m_2 lie on a straight line through O. Their displacements from O are x_1 and x_2 respectively.

The position given by $\bar{x} = \dfrac{m_1 x_1 + m_2 x_2}{m_1 + m_2}$ is called the **centre of mass** of m_1 and m_2.

> If m_1 and m_2 are equal then this result gives the mid-point of the two masses.

For n separate masses m_1, m_2, ... m_n the position of the centre of mass is given by

$$\bar{x} = \frac{m_1 x_1 + m_2 x_2 + ... m_n x_n}{m_1 + m_2 + ... m_n} \quad \text{This is usually written as} \quad \bar{x} = \frac{\displaystyle\sum_{i=1}^{n} m_i x_i}{\displaystyle\sum_{i=1}^{n} m_i}$$

In two dimensions, using the usual Cartesian coordinates, the position of the centre of mass is at $(\bar{x}, \bar{y})$ where $\bar{x}$ and $\bar{y}$ are given by

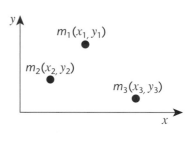

$$\bar{x} = \frac{\displaystyle\sum_{i=1}^{n} m_i x_i}{\displaystyle\sum_{i=1}^{n} m_i} \quad \text{as above, and} \quad \bar{y} = \frac{\displaystyle\sum_{i=1}^{n} m_i y_i}{\displaystyle\sum_{i=1}^{n} m_i}$$

Example

Masses of 2 kg, 3 kg and 5 kg are positioned as shown in the diagram.
Find the coordinates of the centre of mass of the system.

> The method is easily extended to deal with any number of masses.

$$\bar{x} = \frac{2 \times 2 + 3 \times 4 + 5 \times 7}{2 + 3 + 5} = \frac{51}{10} = 5.1$$

$$\bar{y} = \frac{2 \times 1 + 3 \times 5 + 5 \times 3}{2 + 3 + 5} = \frac{32}{10} = 3.2$$

The centre of mass has coordinates (5.1, 3.2)

Centre of mass of a lamina

A **lamina** is something that is thin and flat such as a sheet of metal. The thickness of a lamina is ignored and it is treated as a two-dimensional object.

You can find the centre of mass of a uniform lamina by dividing it into parts which you then represent as particles. Each part is usually rectangular or triangular and in order to represent it as a particle:

- the mass of the particle is represented by the area of the part
- the position of the particle is taken to be at the geometric centre of the part.

This representation of a lamina gives you the information you need to apply the formulae for its centre of mass.

Example

Find the coordinates of the centre of mass of this uniform lamina.

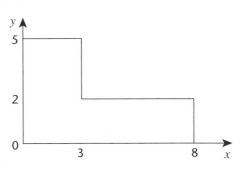

> There are different ways to do this, but it doesn't matter which you choose.

> This process is easily extended to deal with more complex composite figures.

The first step is to divide the lamina into rectangles.

Rectangle A has area 15 square units. Its centre is at (1.5, 2.5).

Rectangle B has area 10 square units. Its centre is at (5.5, 1).

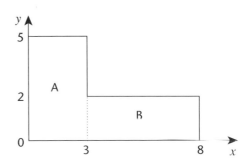

$$\bar{x} = \frac{15 \times 1.5 + 10 \times 5.5}{15 + 10} = 3.1$$

$$\bar{y} = \frac{15 \times 2.5 + 10 \times 1}{15 + 10} = 1.9$$

The centre of mass of the lamina has coordinates (3.1, 1.9).

Freely suspended lamina

Once you know where the centre of mass of a lamina is, you can use this to work out how it will move if it is freely suspended from a given point.

> **KEY POINT**
>
> When a lamina is freely suspended, it will move so that its centre of mass lies directly below the point of suspension.

Example

The diagram shows the lamina of the previous example hanging freely from one corner.

Find the size of angle θ between the direction of Oy and the horizontal.

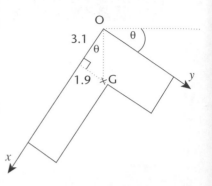

The position of the centre of mass is shown at the point G.
Notice that θ also lies in a right-angled triangle in which two of the sides are known.

From the diagram: $\tan \theta = \dfrac{1.9}{3.1}$

giving $\theta = 31.5°$ to 1 d.p.

Progress check

1 Masses of 2 kg, 3 kg and 5 kg are positioned as shown in the diagram.

   (a) Find the coordinates of G, the centre of mass of the system.
   (b) Find the angle θ between OG and Ox.

1 (a) G(4.5, 2.5) (b) 29.1°

3.5 Uniform circular motion

After studying this section you should be able to:

- *understand angular speed and be able to apply the formula $v = r\omega$*
- *understand that a particle moving in a circular path with constant speed has an acceleration towards the centre of the circle*
- *use the formula $\omega^2 r$, or its equivalent form $\dfrac{v^2}{r}$, to represent the magnitude of the acceleration towards the centre of the circle in solving problems*

LEARNING SUMMARY

Angular speed

Suppose that a particle P moves with constant speed in a circular path with centre O.

The direction of OP changes with time.
The angle θ, through which OP turns in some given time, is usually measured in radians.

The **angular speed** of P is the rate of change of θ with respect to time. Using ω to represent angular speed in radians per second gives:

$$\omega = \frac{\theta}{t}$$

One advantage of using radians is that the relationship between angular speed ω, and linear speed v, is easily expressed as $v = r\omega$.

For example, a particle moving in a circular path of radius 2 m with angular speed 10 radians/sec has a straight line speed of $2 \times 10 = 20$ ms^{-1}.

Radial acceleration

Notice that while the speed of the particle may be constant, its *direction* changes as it moves around the circle. This means that the *velocity* is *not constant* and so the particle must be *accelerating*.

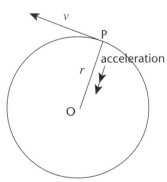

This acceleration is always directed towards the centre of the circle and is known as **radial acceleration** or sometimes **centripetal acceleration**.

The magnitude of this acceleration is given by $a = \omega^2 r$.

Since $v = r\omega$, it follows that $a = \dfrac{v^2}{r}$.

This means that there is a choice of which formula to use. In practice, you should use the one that best fits the available information.

Example

Since we are given the angular speed, the formula $a = \omega^2 r$ is the simpler one to use here.

A particle moves in a circular path of radius 3 m with angular speed 5 radians/sec. Find the magnitude of its radial acceleration.

Using $a = \omega^2 r$: the radial acceleration is $5^2 \times 3 = 75$ ms^{-2}.

Circular motion and force

The radial acceleration of an object moving in a circular path must be produced as the result of a *force acting towards the centre of the circle*. This force is sometimes called the **centripetal force**.

Using $F = ma$, the force needed to keep an object of mass m kg moving in a circular path is $m\omega^2 r$. An alternative form of this is $\dfrac{mv^2}{r}$.

In applying this theory to solving problems, the centripetal force may take various forms such as a tension in a string, friction or the gravitational attraction of a planet.

Example

One end of a string of length 80 cm is attached to a point O on a smooth horizontal table. The other end is attached to a mass of 3 kg moving with speed 4 ms^{-1} and the string is taut. Find the tension in the string.

> It is important to write the length of the string in metres.

Using $T = \dfrac{mv^2}{r}$ gives $T = \dfrac{3 \times 4^2}{0.8} = 60$

The tension in the string is 60 N.

The conical pendulum

One particular situation involving circular motion is described as the **conical pendulum**.

This is where an object moves in a horizontal circle while suspended by a string attached to a point vertically above the centre of the circle.

- The vertical component of the tension in the string supports the weight of the object.
- The horizontal component of the tension in the string provides the centripetal force.

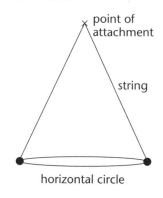

Example

A light inextensible string is fixed at one end to a point O. The other end is attached to a particle of mass 2 kg which moves in a horizontal circle of radius 30 cm with angular speed 4 radians per second. Find the angle of inclination of the string to the vertical. Take $g = 9.8$ ms^{-2}.

Vertically: $\qquad T \cos \theta = 19.6 \qquad\qquad$ (1)

Horizontally: $\quad T \sin \theta = 2 \times 4^2 \times 0.3 = 9.6 \qquad$ (2)

(2) ÷ (1) gives: $\quad \tan \theta = \dfrac{9.6}{19.6} \Rightarrow \theta = 26.1°$

The string is inclined at 26.1° to the vertical.

Progress check

1 One end of a string of length 60 cm is attached to a point O on a smooth horizontal table. The other end is attached to a mass of 5 kg moving with speed 3 ms^{-1} and the string is taut. Find the tension in the string.

3.6 Further circular motion

After studying this section you should be able to:

- *solve problems involving motion in a vertical circle*

LEARNING SUMMARY

Motion in a vertical circle

Solving problems regarding the motion of a particle in a vertical circle generally involves:

- calculation of the speed of the particle by considering energy
- consideration of the radial forces necessary to maintain circular motion.

> For a particle attached to a string it is assumed that the only forces acting on the particle are its weight and the tension in the string. Since the tension always acts in a direction perpendicular to the motion *it does no work* and so the total mechanical energy of the particle is conserved.
>
> **KEY POINT**

Example

A particle P of mass 0.3 kg is attached to one end of a light inelastic string of length 0.8 m. The other end of the string is attached to a fixed point O. The particle is given an initial speed of 5 ms^{-1} horizontally from a point 0.8 m vertically below O. Find:

(a) the speed of P when OP is horizontal
(b) the tension in the string at this point
(c) the angle that OP makes with the upward vertical when the string first becomes slack.

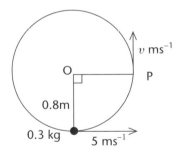

(a) Since energy is conserved:

$$\tfrac{1}{2} \times 0.3 \times v^2 = \tfrac{1}{2} \times 0.3 \times 5^2 - 0.3 \times 9.8 \times 0.8$$

giving: $v^2 = 5^2 - 2 \times 9.8 \times 0.8 = 9.32$

$$\Rightarrow v = 3.052 \ldots$$

so the speed of P when OP is horizontal is 3.05 ms^{-1} to 3 s.f.

$KE = \tfrac{1}{2}mv^2$
$GPE = mgh$

(b) Horizontally: $T = \dfrac{0.3 \times 9.32}{0.8} = 3.495$

The tension in the string is 3.50 N to 3 s.f.

(c) Using the conservation of energy principle:

$$\tfrac{1}{2} \times 0.3 \times v^2 = \tfrac{1}{2} \times 0.3 \times 5^2 - 0.3 \times 9.8(0.8 + 0.8 \cos \theta)$$
So $v^2 = 5^2 - 19.6 \times 0.8(1 + \cos \theta)$ \qquad (1)

When the string first becomes slack, the component of weight along PO is equal to the force needed to maintain circular motion. This gives:

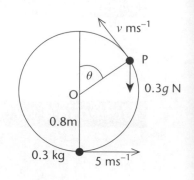

$$0.3 \times 9.8 \cos \theta = \frac{0.3v^2}{0.8} \Rightarrow v^2 = 0.8 \times 9.8 \cos \theta$$

$$(2)$$

From (1) and (2) $\cos \theta = \dfrac{9.32}{23.52} \Rightarrow \theta = 66.7°$

The string first becomes slack when OP makes an angle of 66.7° with the upward vertical.

Progress check

1 A particle P of mass 0.5 kg is attached to one end of a light inextensible string of length 0.6 m. The other end of the string is attached to a fixed point O. P is released from rest with OP horizontal and the string taut.
Find the maximum tension in the string.

1 14.7 N

3.7 Work, energy and power

After studying this section you should be able to:

- *understand kinetic and potential energy and the work energy principle*
- *understand the principle of conservation of mechanical energy*
- *understand the definition of power*
- *solve problems involving work energy and power*

LEARNING SUMMARY

Work and energy

In everyday language, **energy** is needed to do **work**. This idea is given a formal interpretation within mechanics.

The work done, W, by a *constant* force, F, in moving an object through a distance, d, along its line of action is given by $W = Fd$. The unit of work is the **joule** (J).

Energy is also measured in joules and the energy used in moving the object is equal to the work done. In a way, work and energy are like two sides of the same coin.

Energy may take different forms such as heat, sound, electrical, chemical and mechanical. An important property of energy is that it may change from one form to another. For your mechanics module, the focus of attention is on mechanical energy.

There are two forms of mechanical energy: **kinetic energy (KE)**
and **potential energy (PE)**

The kinetic energy of an object is the energy it has due to its *motion*.

Its value depends on the mass of the object and its speed. $KE = \frac{1}{2}mv^2$.

For example, a mass of 4 kg moving with speed 5 ms^{-1} has $KE = \frac{1}{2} \times 4 \times 5^2 = 50$ J

The potential energy of an object is the energy it has due to its *position*.

The **gravitational potential energy** of an object, relative to some level, is equal to the work done against gravity in raising the object from that level to its current position.

PE = mgh relative to ground level.

The total mechanical energy of an object is the sum of its KE and PE.

The **principle of conservation of mechanical energy** states that the total mechanical energy of a system remains constant, provided that it is not acted on by any external force, other than gravity.

KEY POINT

The principle of conservation of energy applies, for example, to the motion of a projectile. Any change in the PE of the projectile corresponds to an equal but opposite change in its KE.

Power

Power is the rate of doing work. It is measured in watts (W) where $1\text{ W} = 1\text{ Js}^{-1}$.

If the point of application of a force F moves with speed v in the *direction of the force* then the power P is given by $P = Fv$.

Example

Find the power generated by a car engine when the car travels at a constant speed of 72 kmh^{-1} on horizontal ground against resistance forces of 2500 N.

> Since the speed is constant, the force produced by the engine must match the resistance force.

$$72\text{ kmh}^{-1} = \frac{72 \times 1000}{60 \times 60}\text{ ms}^{-1} = 20\text{ ms}^{-1}.$$

$$\text{Power} = 2500 \times 20\text{ W} = 50\ 000\text{ W}$$
$$= 50\text{ kW}$$

Exam questions may involve resisted motion on an inclined plane.

Example

A car of mass 1000 kg travels up a hill inclined at angle θ to the horizontal, where $\sin\theta = \frac{1}{20}$. The non-gravitational resistance to motion is 2000 N and the power output from the engine is 60 kW. Find the acceleration of the car when it is travelling at 10 ms^{-1}. Take $g = 10\text{ ms}^{-2}$.

> A clearly labelled diagram is a must.

> In this formula, F represents the *resultant force* up the hill.

Using $P = Fv$: $60\ 000 = F \times 10 \Rightarrow F = 6000$

Using $F = ma$: $6000 - 2000 - 10\ 000 \times \frac{1}{20} = 1000a$

This gives: $3500 = 1000a \Rightarrow a = 3.5$

The acceleration of the car up the hill is 3.5 ms^{-2}.

Progress check

1. A ball is kicked from ground level with speed 15 ms^{-1} and just clears a wall of height 2 m. The speed of the ball as it passes over the wall is $v\text{ ms}^{-1}$.
   (a) Assume that the ball has mass m kg and use the conservation of mechanical energy principle to write an equation.
   (b) Solve the equation to find v.

2. A car of mass 800 kg travels up a hill inclined at angle θ to the horizontal, where $\sin\theta = \frac{1}{20}$. The non-gravitational resistance to motion is 1600 N and the power output from the engine is 45 kW. Find the acceleration of the car when it is travelling at 15 ms^{-1}. Take $g = 10\text{ ms}^{-2}$.

2   1.25 ms^{-2}
(b) 13.6 ms^{-1}
1   (a) $\frac{1}{2}m \times 15^2 = m \times 9.8 \times 2 + \frac{1}{2}mv^2$

3.8 Elastic springs and strings

After studying this section you should be able to:

- *understand and apply Hooke's law, including the term modulus of elasticity*
- *find the energy stored in an elastic string or spring*
- *solve problems involving elastic strings and springs using the work-energy principle*

LEARNING SUMMARY

Hooke's law

An elastic string will vary in length depending on its tension. Its **natural length** l, is its length when the tension is zero. The *extra* length of an elastic string due to a tension T within it is called the **extension** x.

Hooke's law states that the tension in an elastic string is proportional to the extension. This may be expressed as $T \propto x$ or

$$T = kx \qquad (1)$$

for some constant value k.

> When a string has doubled its length, the extension must equal the natural length.

The **modulus of elasticity** λ, is the **tension** required to make an elastic string *double its length*. It follows from (1) that $\lambda = kl$ and $k = \dfrac{\lambda}{l}$.

> Hooke's law is normally used in this form.

Substituting for k in (1): Hooke's law may now be written as $T = \dfrac{\lambda}{l} x$.

Hooke's law applies in the same way to **springs** that are under tension.
The difference is that a spring will also exert an outward force, or thrust, when it is *compressed* by a distance x. The formula given by Hooke's law applies in both situations.

Example

A light elastic string of natural length 0.8 m and modulus of elasticity 30 N is extended to a length of 1.1 m. Find the tension in the string.

In this case:　　　$l = 0.8$, $\lambda = 30$ and $x = 0.3$.

Using Hooke's law gives:　　　$T = \dfrac{30}{0.8} \times 0.3 = 11.25$

The tension in the string is 11.25 N.

Example

A spring of natural length 0.5 m is compressed to a length of 0.35 m by a force of 60 N. Find the modulus of elasticity.

For the spring:　　　$l = 0.5$, $x = 0.15$ and $T = 60$.

Using Hooke's law gives:　　　$60 = \dfrac{\lambda}{0.5} \times 0.15 \implies 0.3\lambda = 60$

so:　　　　　　　　　　　$\lambda = 200$

The modulus of elasticity for the spring is 200 N.

Elastic potential energy

The **elastic potential energy** EPE of an elastic string or spring is its capacity for doing work as it returns to its natural length. Assuming that there is no loss of mechanical energy, this is equivalent to the work done in producing the extension or compression in the first place.

> The area under the graph represents the work done against the tension in producing a given extension.

The diagram shows the linear relationship between T and x.

Work done $= \dfrac{1}{2} \times x \times \dfrac{\lambda}{l} x$

so the EPE $= \dfrac{\lambda x^2}{2l}$

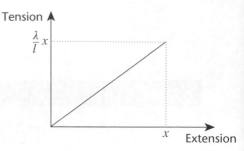

Example

An elastic string of natural length 1 m and modulus of elasticity 30 N is stretched to a length of 1.4 m. Calculate the EPE stored in the string.

> EPE is measured in Joules.

$$\text{EPE} = \dfrac{\lambda x^2}{2l} = \dfrac{30 \times 0.4^2}{2 \times 1} = 2.4 \text{ J.}$$

Example

One end of a light elastic string is attached to a point O, on a smooth horizontal table. The other end is attached to a particle P of mass 0.3 kg held in position on the table such that OP = 1.8 m. The string has modulus of elasticity 20 N and natural length 1.2 m.

The system is released from rest. Find the speed of P when the string becomes slack.

$$\text{EPE} = \dfrac{\lambda x^2}{2l} = \dfrac{20 \times 0.6^2}{2 \times 1.2} = 3 \text{ J.}$$

Since the table is *smooth*, all of the EPE will be converted to KE as the string returns to its natural length and becomes slack.

$\text{KE} = \frac{1}{2}mv^2 = \frac{1}{2} \times 0.3 \times v^2 = 0.15v^2.$

This gives: $0.15v^2 = 3 \implies v = \sqrt{20}$

P has speed 4.47 ms^{-1} when the string becomes slack.

Progress check

1 An elastic string of natural length 0.8 m is extended to a length of 1.4 m by a force of 60 N. Find the modulus of elasticity of the string.
2 A light spring with modulus of elasticity 80 N and natural length 0.9 m is compressed by a distance of 0.2 m. One end of the spring is fixed and a particle of mass 0.1 kg is attached to the other end. The system is released from rest on a smooth horizontal surface.
   Find the maximum speed reached by the particle in the subsequent motion.

2 5.96 ms^{-1} to 3 s.f.
1 80 N

Sample questions and model answers

1

0.5 m

The diagram shows a particle of mass 0.2 kg threaded onto a smooth string fixed at its end-points A and B. The points A, B and C lie in a vertical line and the particle travels in a horizontal circle of radius 0.5 m with centre C.

(a) Calculate the tension in the string.

(b) Calculate the speed of the particle.

(a) Since the string is smooth, the tension in both parts of the string is the same.

> The diagram shows the forces acting on the particle.

40°

T N

80°

T N

0.2 × 9.8 N

The particle is in equilibrium vertically.

Resolving vertically $\qquad T\cos 40° + T\cos 80° = 0.2 \times 9.8$

$$T = \frac{0.2 \times 9.8}{0.93969} = 2.0857\ldots$$

The tension in the string is 2.09 N (3 s.f.)

(b) Let the speed of the particle be v ms^{-1}.

> Centripetal force $= \dfrac{mv^2}{r}$

Resolving horizontally $\qquad$ Centripetal force $= T\sin 40° + T\sin 80°$

$$= 3.3948\ldots = \frac{0.2v^2}{0.5}$$

$$v = \sqrt{\frac{3.3948}{0.4}}$$

$$= 2.913\ldots$$

The speed of the particle is 2.91 ms^{-1} (3 s.f.)

Sample questions and model answers *(continued)*

2

A train of mass 700 tonnes travels at constant speed up a slope inclined at angle $\theta = \sin^{-1}(0.02)$ to the horizontal. The total resistance force is 250 000 N and the power output of the engine is 8600 KW. Take $g = 9.8$ ms^{-2}.

Find the speed, v ms^{-1}, of the train.

It's always a good idea to draw a diagram.

Use one type of arrow for force and another for speed to avoid confusion.

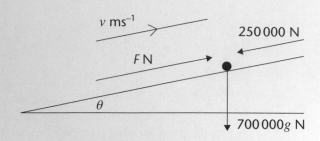

Since the train moves at constant speed, the forces parallel to the slope must balance.

Resolving parallel to the slope

$F = 250\,000 + 700\,000 \times 9.8\sin\theta$

$\quad = 250\,000 + 700\,000 \times 9.8 \times 0.02$

$\quad = 387\,200$

Using $P = Fv$ gives

$8\,600\,000 = 387\,200v$

$v = 22.210\ldots$

The speed of the train is 22.2 ms^{-1} (3 s.f.)

Practice examination questions

1 A particle moves along the x-axis such that at time t seconds its displacement from a point O on the line is $x = t^3 - 6t^2 + 3$ metres.

(a) Show that the particle is initially at rest and find its greatest distance from O in the negative direction.

(b) Find the speed of P when it passes through its initial position.

2 The position vector of a particle P at time t seconds relative to a fixed origin O is given by $\mathbf{r} = (4t^2 + 5)\mathbf{i} + 2t^3\mathbf{j}$ metres.

Find:

(a) the speed of P when $t = 1$

(b) the time when P is moving parallel to the vector $\mathbf{i} + 3\mathbf{j}$.

3 The diagram shows a uniform lamina ABCDEFGH.

(a) Write down the distance of the centre of mass of the lamina from AH.

(b) Find the distance of the centre of mass of the lamina from AB.

(c) The lamina is now suspended freely from A. Find the angle that AB makes with the horizontal in its equilibrium position.

4

The diagram shows a uniform lamina of mass 0.8 kg resting on supports at A and B.

(a) Find the horizontal distance of the centre of mass from the support at A.

(b) Find the magnitude of the reaction force on the lamina at each support.

5 A motorbike is driven round a corner in a circular path of radius 40 m. The maximum speed that this can be done without skidding is 15 ms^{-1}.

Calculate the coefficient of friction between the tyres and the road surface to 2 s.f.

Take $g = 9.8$ ms^{-2}.

6 One end of a light inelastic string of length 50 cm is attached to a fixed point 40 cm above a smooth horizontal table. The other end is attached to an object of mass 2 kg moving in a circular path of radius 30 cm on the table.

Given that the angular speed of the object is 3 radians per second, find:

(a) the magnitude of the acceleration of the object

(b) the tension in the string

(c) the reaction force of the table on the object.

7 A particle P of mass m is attached to one end of a light inelastic string of length l. The other end of the string is attached to a fixed point O. P is released from rest with the string taut and horizontal.

(a) Explain why, in the subsequent motion, the tension in the string does no work on the particle.

(b) Explain why it is not valid to use the formula $v^2 = u^2 + 2as$ in this situation.

(c) Find the maximum speed of the particle assuming that there is no resistance to motion.

8 A car of mass 900 kg travels up a straight road inclined at angle α to the horizontal, where $\sin \alpha = \frac{1}{15}$. Assume that the total non-gravitational force opposing the car's motion has a constant value of 800 N. Take $g = 9.8$ ms^{-2}.

(a) Find the driving force required for the car to maintain a steady speed up the slope.

(b) Find the power output of the car's engine given that its speed up the slope has a constant value of 20 ms^{-1}.

(c) At the top of the hill, the road surface is horizontal. If the power output of the engine remains the same, what is the initial acceleration of the car?

9 A particle P moves along a straight line passing through the point O. The displacement x metres, of the particle from O at time t seconds is given by:

$x = 4t^3 - 3t^2 + 7$ for $0 \leqslant t \leqslant 2$.

Find:

(a) the velocity of the particle at time t

(b) the distance that the particle travels before it changes direction

(c) the time taken for the particle to return to its starting point.

10 A particle moving in a straight line is acted on by a single retarding force with magnitude proportional to the square of its speed. The initial speed of the particle is 20 ms^{-1} and its initial retardation is 3 ms^{-2}.

(a) Show that the speed v ms^{-1} of the particle after t seconds is given by:

$$v = \frac{400}{3t + 20}$$

(b) Given that the particle has mass 2 kg, find the magnitude of the work done by the retarding force between $t = 0$ and $t = 4$.

Practice examination questions

11 A light elastic string of natural length 2 m and modulus of elasticity $5mg$ has one end fixed at a point O. The other end is attached to an object of mass m which is released from rest at O.

Find:

(a) the distance that the object falls below O before being brought to instantaneous rest by the tension in the string

(b) the acceleration of the object at its lowest point.

12 The diagram shows a particle of mass 0.2 kg attached to one end of a light inelastic string of length 80 cm.

The particle moves in a horizontal circle on a smooth surface with the string inclined at 30° to the vertical.

(a) Given that the speed of the particle is 1.2 ms^{-1} find the tension in the string and the reaction of the table on the particle.

(b) Find the minimum angular speed of the particle so that the reaction of the table on the particle is zero.

13 A small ball B is released from rest at the top of a horizontal circular tube. The ball travels down the outside of the tube's surface following a circular path, with centre C, until it falls away.

Given that the outside radius of the tube is 0.6 m and the mass of the ball is 0.1 kg find:

(a) an expression for the speed of B when CB makes an angle θ with the upward vertical

(b) the reaction of the tube on the ball in terms of θ and g

(c) the value of θ when the ball loses contact with the surface of the tube.
Take $g = 9.8$ ms^{-2}.

Chapter 4

Decision Mathematics 2

The following topics are covered in this chapter:

- *Game theory*
- *Networks*
- *Critical path analysis*

- *Linear programming*
- *Matching and allocation*
- *Dynamic programming*

4.1 Game theory

LEARNING SUMMARY

After studying this section you should be able to:

- *understand the idea of a zero-sum game and its representation using a pay-off matrix*
- *identify play-safe strategies and stable solutions*
- *find optimal mixed strategies for a game with no stable solution*
- *reduce a pay-off matrix using a dominance argument*

A two person zero-sum game

The starting point in the mathematical theory of games is that the outcome of a game is determined by the **strategies** of the players.

A two person **zero-sum game** is a game in which the winnings of one player equal the losses of the other for every combination of strategies. Taking winnings to be positive and losses to be negative gives a zero sum in each case.

Viewing a game from one player's point of view, we could represent the outcomes (called pay-offs) for each combination of strategies in a matrix. This is called the **pay-off matrix** for that player.

Example

A and B are two players in a zero-sum game. A uses one of two strategies, W or X, and B uses one of the strategies Y or Z. The table shows the pay-off matrix for A.

> The situation could equally be represented by the pay-off matrix for B. This would show corresponding values with opposite signs since this is a zero-sum game.

Pay-off matrix for A		B	
		Y	Z
A	W	2	−2
	X	5	−4

The pay-off matrix shows that if B adopts strategy Y then the pay-off for A will be 2 by using strategy W and 5 using strategy X. On the other hand, if B adopts strategy Z then the pay-off for A will be −2 using strategy W and −4 using strategy X.

The idea is that neither player knows in advance which strategy the other will use.

The play-safe strategy

The **play-safe** strategy for a player is the strategy for which the minimum pay-off is as high as possible. In the example above, the minimum pay-off for A using strategy W is −2, whereas the minimum pay-off using strategy X is −4. This means that strategy W is the play-safe strategy for player A.

Notice that finding the play-safe strategy for player A involves comparing the minimum values in the *rows* of the pay-off matrix for A. Finding the play-safe strategy for B will involve comparing the values in the *columns*, remembering that B's pay-offs are the negatives of the ones in the pay-off matrix for A.

The minimum value for B using strategy Y is −5 and the minimum value using strategy Z is 2. This means that the play-safe strategy for B is strategy Z.

The situation is shown in the pay-off matrices for A and B.

Pay-off matrix for A		B	
		Y	Z
A	W	2	−2
	X	5	−4

Pay-off matrix for B		B	
		Y	Z
A	W	−2	2
	X	−5	4

In this case, the maximum of the minimum pay-offs, for each player, is in the corresponding position in the two matrices. This represents the **stable solution** to the problem referred to as the **saddle point** (or **minimax point**).

The solution is stable in the sense that neither player can improve their pay-off by taking a different strategy, given that the other player doesn't change. In other words, while B uses strategy Z, the best strategy for A is W and while A uses strategy W, the best strategy for B is Z.

In the example, the sum of these values is −2 + 2 = 0 and the solution is stable.

> If the sum of the two values used to determine the play-safe strategies is **not zero** then the values cannot correspond to the same cell position in the play-off matrices. This means that there is no saddle point and the game has **no stable solution**.
>
> **KEY POINT**

Example

The pay-off matrices for two players in a zero-sum game are given below. Show that there is no stable solution for the game.

Pay-off matrix for P		Q	
		Y	Z
P	W	5	−3
	X	−4	6

Pay-off matrix for Q		Q	
		Y	Z
P	W	−5	3
	X	4	−6

The play-safe strategies for P and Q are shown shaded.
The values used to determine the play-safe strategies are −3 and −5.
There is no stable solution since $(-3) + (-5) \neq 0$.

Optimal strategies for games that are not stable

The repeated use of the same strategy over a series of games is called a **pure strategy**. This provides the best results for both players in a game which has a stable solution. In the case where no stable solution exists, a **mixed strategy** is used in which each of the strategies is employed with a given probability to find the optimal solution.

Returning to the previous example:

Pay-off matrix for P	Q Y	Z
P W	5	−3
P X	−4	6

Suppose that P chooses:

strategy W with probability p

and   strategy X with probability $(1 - p)$.

Then if:

Q chooses strategy Y, the expected *gain* for P is given by $5p - 4(1 - p) = 9p - 4$

Q chooses strategy Z, the expected *gain* for P is $-3p + 6(1 - p) = -9p + 6$

> The optimal value occurs when these expressions are equal.

$$9p - 4 = -9p + 6 \Rightarrow 18p = 10$$
$$\Rightarrow p = \tfrac{5}{9}$$

> You could equally use $-9p + 6$ to get the value of the game.

The value of the game is given by $9p - 4 = 1$.

Suppose that Q chooses:

strategy Y with probability q

and   strategy Z with probability $(1 - q)$.

Then if:

P chooses strategy W, the expected *loss* for Q is given by $5q - 3(1 - q) = 8q - 3$

P chooses strategy X, the expected *loss* for Q is $-4q + 6(1 - q) = -10q + 6$

$$8q - 3 = -10q + 6 \Rightarrow 18q = 9$$
$$\Rightarrow q = \tfrac{1}{2}$$

(As a check, the value of the game is $8q - 3 = 1$ as before).

This shows that the optimal strategy for both players is to use mixed strategies such that:

P chooses strategy W with probability $\tfrac{5}{9}$ and strategy X with probability $\tfrac{4}{9}$

Q chooses strategy Y and strategy Z with equal likelihood.

The expected long-term gain then for P, as an average per game, is 1 and this is also the expected long-term loss, as an average per game, for Q.

Graphical representation

Each graph corresponds to a strategy for Q (i.e. the opponent of P).

The diagram shows graphs of $9p - 4$ and $-9p + 6$ against values of p from 0 to 1.

The point of intersection corresponds to the probability that gives the optimal mixed strategy for P in the last example.

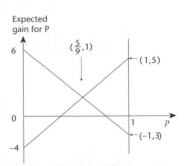

> **KEY POINT**
>
> When P's opponent has more strategies there will be more graphs with several points of intersection. You will need to identify the one that represents the optimal mixed strategy for P.
> *This will be the highest point on or below each of the graphs.*

This diagram represents a situation where P's opponent has three strategies to choose from.

The point representing the optimal mixed strategy for P is circled.

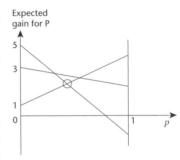

Notice how the problem of identifying the right vertex can be expressed as a linear programming problem in which the object is to maximise the expected gain for P subject to the constraints represented by the regions bounded by the straight line graphs.

This is, in fact, the approach used for higher dimensional problems. The conditions are formulated as a linear programming problem which may then be solved by the Simplex algorithm.

> **KEY POINT**
>
> If the pay-offs for one strategy are *always* better than the corresponding pay-offs for some other strategy then the weaker one can be ignored when determining the probabilities for a mixed strategy. In this way, the pay-off matrix is reduced by what is called a **dominance argument**.

Progress check

1 The table shows a pay-off matrix for player A in a zero-sum game.

   (a) Find the play-safe strategy for each player.
   (b) Show that this game has no stable solution.
   (c) Find the best strategy for each player and the value of the game.

Pay-off matrix for A		Q	
		Y	Z
A	W	6	-4
	X	-3	5

1 (a) A(strategy X), Q(strategy Z)
(b) $(-3) + (-5) \neq 0$
(c) A Use W and X with probabilities $\frac{4}{9}$ and $\frac{5}{9}$
Q Use Y and Z with equal probability, value of game = 1

4.2 Networks

After studying this section you should be able to:

- *understand flows in networks and be able to find the maximum flow for a network involving multiple sources and sinks*

LEARNING SUMMARY

Flows in networks

The **flows** referred to may be flows of liquids, gases or any other measurable quantities. The edges may represent such things as pipes, wires or roads that carry the quantities between the points identified as vertices.

A typical vertex has a flow into it and a flow out of it. The exceptions are a **source** vertex which has no input and a **sink** vertex which has no output.

Each edge of the network has a **capacity** which represents the maximum possible flow along that edge. In the usual notation, the capacity is written next to the edge and the flow is shown in a circle.

This edge has a capacity of 10 and a flow of 7.

The set of flows for a network is **feasible** if:

- The total output from all source vertices is equal to the total input for all sink vertices.

- The input for each vertex other than a source or sink vertex is equal to its output.

- The flow along each edge is less than or equal to its capacity.

A **cut** divides the vertices into two sets, one set containing the source and the other containing the sink.

The **capacity of a cut** is equal to the sum of the capacities of the edges that cross the cut, *taken in the direction from the source set to the sink set*.
In the diagram, the capacity of cut (i) is 15 + 10 = 25.
The capacity of cut (ii) is 8 + 14 = 22.
Notice that, for the second cut, the capacity of 4 is not included because it does not go from the source set to the sink set.

> Be careful to add the capacities and not the flows.

The maximum flow–minimum cut theorem

The **maximum flow–minimum cut theorem** states that the maximum value of the flow through a network is equal to the capacity of the minimum cut. (This is a little bit like saying that a chain is only as strong as its weakest link.)

It follows from the theorem that if you find a flow that is equal to the capacity of some cut then it must be the maximum flow that can be established through the network.

For example, the capacity of cut (ii) for this network has already been found to be 22. The circled figures show one way of achieving this flow and the theorem now tells us that this must be the maximum possible flow through the network.

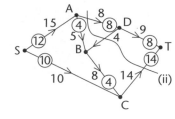

Flow augmentation

Once an initial flow has been established through a network it is useful to know the extent to which the flow along any edge may be altered in either direction.

This is labelled as **excess capacity** and **back capacity** on each edge as shown. It may be possible to **augment** the flow through the network by making use of this flexibility along a path from the source vertex to the sink vertex. Such a path is called a **flow-augmenting path**.

A **saturated** edge is one where the flow is equal to the capacity.

An algorithm for finding the maximum flow through a network by augmentation is:

Step 1   Find an initial flow by inspection.

Step 2   Label the excess capacity and back capacity for each edge.

Step 3   Search for a flow-augmenting path. If one can be found then increase the flow along the path by the maximum amount that remains feasible.

Step 4   Repeat steps 2 and 3 until no flow-augmenting path may be found.

Multiple sources and sinks

A network may have several sources S_1, S_2, ... and several sinks T_1, T_2, ... but the methods described in this section can still be applied through the use of a **supersource** and a **supersink**. A supersource S is an extra node, added to the network, that acts as a source for each of the original sources. Similarly, a supersink T is a node, added to the network, that acts as a sink for each of the original sinks.

Example

The network below shows two sources S_1 and S_2 and two sinks T_1 and T_2. Find the maximum flow through the network and state any flow-augmenting paths.

Explain why the flow is maximal.

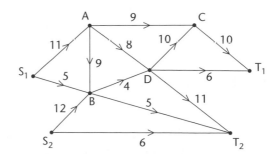

The first step is to add the supersource S and the supersink T to the network.

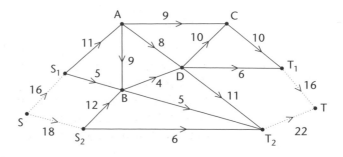

The usual convention is to use dotted lines for the edges linking S and T to the network.

The capacities on the edges SS_1 and SS_2 have been chosen to support the maximum possible flows from S_1 and S_2. Similarly, the capacities on the edges T_1T and T_2T have been chosen to support the maximum flows into T_1 and T_2.
The diagram also shows a cut with capacity $11 + 4 + 5 + 6 = 26$.

A flow of 26 can be achieved as shown in the table.

Flow-augmenting path	Flow	Saturated edges
SS_1ACT_1T	9	S_1A, AC
SS_1ADT_1T	2	S_1A, DT_1
SS_2BDT_1T	4	BD, DT_1
SS_2BT_2T	5	BT_2
SS_2T_2T	6	S_2T_2
Total	26	

The flows on the flow-augmenting paths are shown circled on the network below. The flow of 26 is maximal since it equals the capacity of the cut (maximum flow-minimum cut theorem).

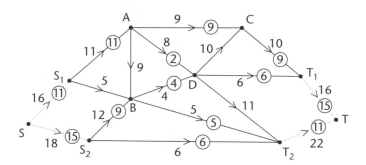

Note that each of the edges on the minimum cut is saturated. This will always be the case.

The final stage is to remove the supersource and supersink, along with the extra arcs, to show the maximum flow pattern for the original network.

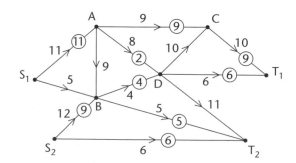

Progress check

1 Find the maximum flow through this network. Verify your answer using the maximum flow-minimum cut theorem.

2 Use a supersource and supersink to find the maximum flow through this network.
  Establish the flow through B in this case.

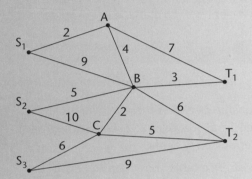

4.3 Critical path analysis

After studying this section you should be able to:

- *construct an activity network from a given precedence table*
- *use forward and backward scans to determine the earliest start times and the latest finish times*
- *find the critical path in an activity network*
- *calculate the total float of an activity*
- *construct a chart for the purpose of scheduling*
- *draw a resource histogram and use resource levelling*

LEARNING SUMMARY

The process of representing a complex project by a network and using it to identify the most efficient way to manage its completion is called **critical path analysis**.

Activity networks

A complex project may be divided into a number of smaller parts called **activities**.

The relationship between these activities can be represented in a **precedence table**, sometimes called a **dependency table**.

In the precedence table shown on the right, the figures in brackets represent the **duration** of each activity, i.e. the time required, in hours, for its completion.

A precedence table can be used to produce an **activity network**. Each activity is represented by a node labelled as below.

Activity	Depends on
A(3)	–
B(5)	–
C(2)	A
D(3)	A
E(3)	B, D
F(5)	C, E
G(1)	C
H(2)	F, G

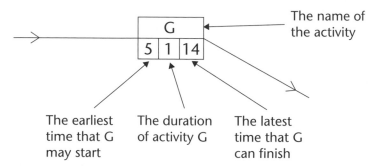

The name of the activity

The earliest time that G may start

The duration of activity G

The latest time that G can finish

Precedence within the network is indicated by linking the activities with arrows.

Earliest start times

The **earliest start time** for an activity is the earliest time by which all dependent events may be completed. These times are calculated using a **forward pass**, or **forward scan**, starting from the first activity and working towards the last.

Latest finish times

The **latest finish time** for an activity is the latest time by which the activity may be completed without extending the time for the project. These times are calculated using a **backward pass**, or **backward scan**, starting from the last activity and working back through the network.

A **critical activity** is one whose duration cannot be increased without extending the time for the project. At a critical activity, the earliest start time plus the duration gives the latest finish time. A path through the network made entirely of critical activities is a **critical path**. Some networks have more than one critical path.

It is useful to add the information about earliest and latest times to the network. The critical path is then easily identified.

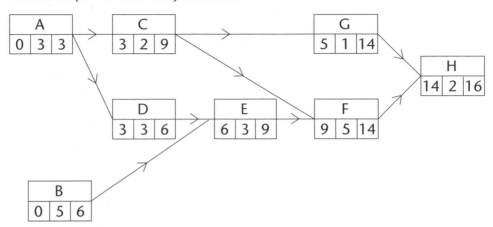

In this case, the critical path is ADEFH.

The **total float of an activity** is the maximum time that the activity may be delayed without affecting the length of the critical path. It is given by:

latest finish time – earliest start time – duration of the activity.

> The total float of any critical activity is always zero.

Scheduling

The process of allocating activities to workers for completion, within all of the constraints of the project, is known as **scheduling**.

The information regarding earliest and latest times for each activity is crucial when constructing a schedule. This information may be presented as a table or as a chart.

> Typically, the purpose of scheduling is to determine the number of workers needed to complete the project in a given time, or to determine the minimum time required for a given number of workers to complete the project.

		Start		Finish		
Activity	Duration	Earliest	Latest	Earliest	Latest	Float
A(0, 1)	3	0	0	3	3	0
B(0, 2)	5	0	1	5	6	1
C(1, 3)	2	3	7	5	9	4
D(1, 2)	3	3	3	6	6	0
E(2, 4)	3	6	6	9	9	0
F(4, 5)	5	9	9	14	14	0
G(3, 5)	1	5	13	6	14	8
H(5, 6)	2	14	14	16	16	0

The critical activities are shown along one line.

The diagram illustrates the degree of flexibility in starting activities B, C and G. Remember that G cannot be started until C has been completed.

> The diagram shown here is known as a cascade chart or Gantt chart.

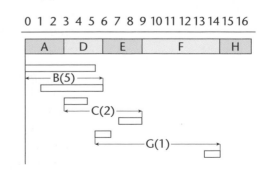

Resource histograms and resource levelling

A **resource histogram** shows the demand on resources over the time taken to complete a project.

The example below shows how the number of workers needed varies from day to day over a period of 14 days.

The maximum number of workers needed is 6 for days 7 and 8.

The diagram below shows the same resource histogram but includes details of the duration for each activity and the number of workers. Activity F, for example, lasts for 3 days and requires 2 workers. The critical activities are A, C, D, E and G. These are put in first and the rest of the diagram is built in layers on top.

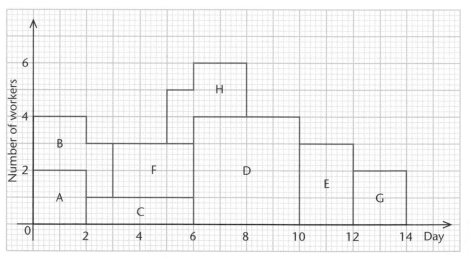

Once the resource histogram has been drawn, you may be asked to use **resource levelling** to work out the effect on the schedule of having a reduction in resources. For example, if the maximum number of workers available at any time is 5 then we can see from the diagram that this presents a problem for the completion of activity H.

Resource levelling involves selecting one or more activities to be delayed by an amount of time that allows the demand on resources to be brought down to the required level. The consequences of each delay must then be taken into account to determine the amount of extra time required.

In this example, delaying the start of activity D by 2 days allows activity H to be completed. Activities E and G must then also be delayed by 2 days and so the whole project requires 2 more days for completion.

Progress check

1 Here is the activity network used in the example on page 113.

	A				C				G		
0	3	3		3	2	9		5	1	14

	H	
14	2	16

	D				E				F		
3	3	6		6	3	9		9	5	14

	B	
0	5	6

The number of workers required for each activity is shown in the table.

A	B	C	D	E	F	G	H
2	3	4	1	2	3	2	2

Assume that all workers are equally skilled and that once an activity has been started it must be completed without a break. Assume also that each activity starts at its earliest time.

(a) Draw the resource histogram for the project assuming that the number of workers available is unlimited.

(b) Given that the maximum number of workers available at any time is 5, use resource levelling to find the extra time needed to complete the project.

(b) Delay the start of C by 2 days.
Delay the start of G by 2 days.
Delay the start of E, F and H by 1 day.
The extra time needed is 1 day.

1 (a)

4.4 Linear programming

After studying this section you should be able to:

- *use the Simplex algorithm to solve a problem algebraically*

LEARNING SUMMARY

The Simplex algorithm

When there are three or more variables, a different approach is needed. The **Simplex algorithm** may be used to solve the problem algebraically. The information must be expressed in the right form before the algorithm can be used.

- First write the constraints, other than the non-negativity conditions, in the form $ax + by + cz \leqslant d$.
- Write the objective function in a form which is to be maximised.
- Add **slack variables** to convert the inequalities into equations.
- Write the information in a table called the **initial tableau**.

> The objective function is already in a form which is to be maximised.

> The constraints are in the right form.

Example

Find the maximum value of $P = 2x + 3y + 4z$ subject to the constraints:

$$3x + 2y + z \leqslant 10 \tag{i}$$
$$2x + 5y + 3z \leqslant 15 \tag{ii}$$
$$x \geqslant 0, \quad y \geqslant 0, z \geqslant 0.$$

Using slack variables s and t, the inequalities [i] and [ii] become:

$$3x + 2y + z + s = 10$$
$$2x + 5y + 3z + t = 15.$$

The objective function must be rearranged so that all of the terms are on one side of the equation to give:

$$P - 2x - 3y - 4z = 0.$$

The information may now be put into the initial tableau.

	P	x	y	z	s	t	value
The top row shows the objective function	1	−2	−3	−4	0	0	0
This row shows the first constraint	0	3	2	1	1	0	10
This row shows the second constraint	0	2	5	3	0	1	15

The columns for P, s and t contain zeros in every row apart from one. In each case, the remaining row contains 1 and the value of the variable is given in the end column of that row. The value of every other variable is taken to be zero. So, at this stage, the tableau shows that $P = 0$, $s = 10$, $t = 15$ and x, y and z are all zero. This corresponds to the situation at the origin.

> You need to know how to read values from the tableau.

Using the algorithm is equivalent to visiting each vertex of the feasible region until an optimum solution is found. This occurs when there are no negative values in the objective row.

When the present tableau is not optimal, a new tableau is formed as follows:

- The column containing the most negative value in the objective row becomes the **pivotal column**. In this case the pivotal column corresponds to the variable z.

- Now divide each entry in the *value* column by the corresponding entry in the pivotal column provided that the pivotal column entry is positive.
  For the example above this gives $\frac{10}{1} = 10$ and $\frac{15}{3} = 5$
  The smallest of these results relates to the bottom row which is now taken to be the **pivotal row**. The entry lying in both the pivotal column and the pivotal row then becomes the pivot. In this case, the pivot is 3.

P	x	y	z	s	t	value	
1	−2	−3	−4	0	0	0	
0	3	2	1	1	0	10	
0	2	5	3	0	1	15	← pivotal row

pivotal column

- Divide every value in the pivotal row by the pivot, this makes the value of the pivot 1. The convention is to use fraction notation rather than decimals.

P	x	y	z	s	t	value	
1	−2	−3	−4	0	0	0	
0	3	2	1	1	0	10	
0	$\frac{2}{3}$	$\frac{5}{3}$	1	0	$\frac{1}{3}$	5	← pivotal row

- Now turn the other values in the pivotal column into zeros by adding or subtracting multiples of the pivotal row.

See paragraph on preceding page about reading values from the tableau.

P	x	y	z	s	t	value	
1	$\frac{2}{3}$	$\frac{11}{3}$	0	0	$\frac{4}{3}$	20	
0	$\frac{7}{3}$	$\frac{1}{3}$	0	1	$-\frac{1}{3}$	5	
0	$\frac{2}{3}$	$\frac{5}{3}$	1	0	$\frac{1}{3}$	5	← pivotal row

It is often necessary to repeat the process in order to find the optimum tableau.

- If there are no negative values in the objective row then the **optimum tableau** has been found. Otherwise choose a new pivotal column and repeat the process.

Each time through the process is called an **iteration**.

In this case the optimum tableau has been found after just one iteration. It shows that the maximum value of P is 20 and that this occurs when

$x = 0$, $y = 0$, $z = 5$, $s = 5$ and $t = 0$.

Progress check

An initial simplex tableau for a linear programming problem is given by:

P	x	y	z	s	t	value
1	-3	-2	-1	0	0	0
0	2	3	1	1	0	10
0	3	4	2	0	1	18

1 Carry out one iteration to produce the next tableau.

2 Explain why the tableau found is optimal. Write down the maximum value of P and the corresponding values of the other variables.

2 There are no negative values in the objective row.
$P = 15$, $x = 5$, $y = 0$, $z = 0$, $s = 0$, $t = 3$.

P	x	y	z	s	t	value
1	0	$\frac{5}{2}$	$\frac{1}{2}$	$\frac{3}{2}$	0	15
0	1	$\frac{3}{2}$	$\frac{1}{2}$	$\frac{1}{2}$	0	5
0	0	$-\frac{1}{2}$	$\frac{1}{2}$	$-\frac{3}{2}$	1	3

4.5 Matching and allocation

After studying this section you should be able to:

- *understand the conditions for matchings to be maximal or complete*
- *apply the maximum matching algorithm*

LEARNING SUMMARY

The matching improvement algorithm

Allocation problems

In an allocation (or assignment) problem, the object is to set up a matching that will optimise a particular objective, such as minimising a cost or maximising a profit. If we regard one set of objects as agents and the other set as tasks, then the underlying assumptions are that:

- Any agent may be assigned to any task (unlike the matching problems).
- Each agent is assigned at most one task and each task has at most one agent assigned to it.

When the number of agents is equal to the number of tasks, the problem is said to be **balanced**. If the number of agents exceeds the number of tasks, or vice versa, then the problem is **unbalanced**. The technique for solving an unbalanced problem involves introducing a **dummy** agent or task which then allows the solution to proceed as for the balanced case. In the solution of an unbalanced problem, a task assigned to a dummy agent isn't actually completed and an agent assigned to a dummy task doesn't actually have a task to carry out.

The opportunity cost matrix

The table shows the time in hours for three workers A, B and C to complete three separate tasks.

		Task		
		1	2	3
Worker	A	9	11	7
	B	8	10	9
	C	7	9	8

The **cost matrix** shows only the figures in hours. In this case, the 'cost' is measured in time.

9	11	7
8	10	9
7	9	8

The **opportunity cost matrix** is found from the cost matrix by:

1. Subtracting the smallest value in each row from every other value in that row.
2. Subtracting the smallest value in each column from every other value in that column.

9	11	7
8	10	9
7	9	8

$\xrightarrow{1}$

2	4	0
0	2	1
0	2	1

$\downarrow 2$

2	2	0
0	0	1
0	0	1

The Hungarian algorithm

The **Hungarian algorithm** is used to find the allocation that optimises the given objective.

Step 1 Find the opportunity cost matrix.

Step 2 Test for optimality. If true then make the allocation and stop.

Step 3 Revise the opportunity cost matrix and repeat step 2.

The method used to test for optimality is to find the minimum number of horizontal and/or vertical lines needed to cover all of the zeros in the opportunity cost matrix. If this number $= n$ where the matrix has n rows and n columns, then the matrix will give an optimal allocation.

In the example, the number of lines needed $= 3$.
The matrix has 3 rows and columns so it will give an optimal allocation.

The optimal allocation is now found by selecting positions of zeros in the opportunity cost matrix such that one zero is selected in each row and in each column.

One possibility is shown here along with the corresponding figures in the original cost matrix.

2	2	**0**
0	**0**	1
0	0	1

9	11	**7**
8	**10**	9
7	9	8

The allocation is A $\rightarrow$ 3, B $\rightarrow$ 2 and C $\rightarrow$ 1 giving a total cost, in hours, of $7 + 10 + 7 = 24$.

In this case, there is an alternative solution.

2	2	**0**
0	0	1
0	**0**	1

9	11	**7**
8	10	9
7	**9**	8

The allocation is A $\rightarrow$ 3, B $\rightarrow$ 1 and C $\rightarrow$ 2 giving a total cost, in hours, of $7 + 8 + 9 = 24$.

Revising the opportunity cost matrix

When the opportunity cost matrix does not give an optimal solution, a revised opportunity cost matrix is found as follows:

Step 1 Find the smallest number not covered by a line. Subtract this number from every number not covered by a line.

Step 2 Add this number to every number lying at the intersection of two lines.

For example, this opportunity cost matrix has 3 rows and columns, but its zeros can be covered with just 2 lines.

0	3	1
0	2	5
0	0	0

Applying the steps given above produces this revised opportunity cost matrix.

0	2	**0**
0	1	4
1	**0**	0

Maximising allocation problems

If the objective involves maximising a quantity, such as a profit, then the first step is to subtract every value in the original matrix from a fixed value. This fixed value may be taken to be the largest value in the matrix or any value larger than this. The Hungarian algorithm is now applied in the usual way to this transformed matrix.

Progress check

1 Given the cost matrix:

	1	2	3
A	3	3	4
B	8	5	9
C	2	1	4

Use the Hungarian algorithm to find the minimum total cost.

1 4 + 5 + 2 = 11

4.6 Dynamic programming

After studying this section you should be able to:

- use dynamic programming to optimise a quantity
- solve maximin and minimax problems

LEARNING SUMMARY

Dynamic programming

Dynamic programming is one of the techniques used to solve problems where some quantity is to be optimised. It involves identifying **stages** and all possible **states** and **actions** within a process. The stages are counted from the end of the process to the beginning.

Working backwards through the stages, decisions are made at each stage that optimise the quantity. Importantly, the calculations carried out must follow from the corresponding optimal value found at the previous stage. In effect, this is applying the **principle of optimality** which states that **any part of an optimal path is itself optimal**.

In the examination, information about the process and the quantity to be optimised may be given in the form of a network or a table.

Example

The network below shows the possible actions and corresponding anticipated profits for a company over the next five months. The figures represent the profit in units of ten thousand pounds, with a negative value indicating a loss due to investment.
(a) Use dynamic programming to maximise the expected profit over the five months.
(b) Give the sequence of vertices in order to achieve the maximum profit.

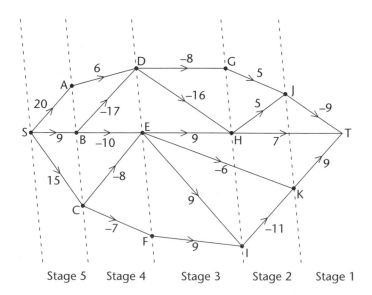

Again, in the examination, an insert is usually provided showing a table to be completed. The table on the next page is typical of the ones used and the coloured entries are typical of the information that would be given on the insert.

(a) An asterisk in the right hand column is used to indicate the optimal value from the given state.

Stage	State	Action	Value	
1	J	JT	−9	*
	K	KT	9	*
2	G	GJ	5 + −9 = −4	*
	H	HJ	5 + −9 = −4	
		HT	7	*
	I	IK	−11 + 9 = −2	*
3	D	DG	−8 + −4 = −12	
		DH	−16 + 7 = −9	*
	E	EH	9 + 7 = 16	*
		EK	−6 + 9 = 3	
		EI	9 + −2 = 7	
	F	FI	9 + −2 = 7	*
4	A	AD	6 + -9 = −3	*
	B	BD	−17 + −9 = -26	
		BE	−10 + 16 = 6	*
	C	CE	−8 + 16 = 8	*
		CF	−7 + 7 = 0	
5	S	SA	20 + −3 = 17	
		SB	9 + 6 = 15	
		SC	15 + 8 = 23	*

JT and KT are the only possible paths from J and K, so they are automatically optimal.

The value for GJ is calculated as 5 (shown on GJ) plus −9 (the optimal value from J). This is the only value from G so it is automatically optimal.

There are two possible moves from H. The optimal value is 7.

There are three possible moves from E. The value for EH, for example, is 9 (shown on EH) plus 7 (the optimum value for H). This gives 16, which turns out to be the optimal value from E.

F is the last vertex to consider from Stage 3.

The three vertices to consider in Stage 4 are A, B and C.

The highest value found at the final stage is the overall maximum value. In this case this is 23 which represents 23 × £10 000.

The maximum expected profit over the next five months is £230 000.

(b) The sequence of vertices that maximises the profit is SCEHT.

Different types of problem

The dynamic programming technique may be used to solve various types of problem. The way in which the table is used remains the same, but the way in which the optimum value is selected may be different. To find the minimum distance through a network, for example, select the minimum value for each state rather than the maximum as in the example above.

In particular, two types that you need to be aware of are the **maximin** and **minimax**.

The optimal values for the maximin are found by first selecting the minimum value for each action and then choosing the maximum of these to represent the state.

To illustrate how the maximin works, here is part of a network and the corresponding part of a table.

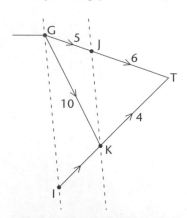

JT and KT are the only possible paths from J and K, so they are automatically optimal.

The value for GJ is the minimum of 5 (shown on the arc GJ) and 6 (the optimum from J at the previous stage).

The value for GK is the minimum of 10 (shown on the arc GJ) and 4 (the optimum from K at the previous stage).

There are two values for the state G and the maximum of these (5) is shown with an asterisk as this is the optimum for G.

Stage	State	Action	Value	
1	J	JT	6	*
	K	KT	4	*
2	G	GJ	Min(5, 6) = 5	*
		GK	Min(10, 4) = 4	

In a similar way, the optimal values for the minimax are found by first selecting the maximum value for each action and then choosing the minimum of these to represent the state.

The table shows the minimax values for the part of the network given above.

Stage	State	Action	Value	
1	J	JT	6	*
	K	KT	4	*
2	G	GJ	Max(5, 6) = 6	*
		GK	Max(10, 4) = 10	

Progress check

1 The diagram shows part of a network. Find the maximin value for the initial state A.

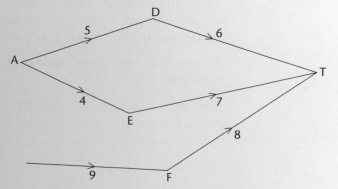

2 Find the minimax value for the initial state A in the diagram in question 1.

2 6
1 5

Sample questions and model answers

1

Ashley and Emma play a two-person zero-sum game in which they each consider three possible strategies. The table shows the pay-off matrix for Ashley.

		Emma		
		X	Y	Z
	A	5	−3	3
Ashley	B	4	1	−2
	C	−3	6	−4

(a) Find the play-safe strategy for each player and show that the game does not have a stable solution.

(b) Set up a linear programming problem to find Ashley's optimal mixed strategy. You are not required to solve it.

(a) The minimum row values are −3, −2 and −4 in Ashley's pay-off matrix. The maximum of these is −2 and so the play-safe strategy for Ashley is strategy B.

The minimum column values in Emma's pay-off matrix are −5, −6 and −3. The maximum of these is −3 and so the play-safe strategy for Emma is strategy Z.

		Emma		
		X	Y	Z
	A	5	−3	3
Ashley	B	4	1	−2
	C	−3	6	−4

		Emma		
		X	Y	Z
	A	−5	3	−3
Ashley	B	−4	−1	2
	C	3	−6	4

The values used to determine the play-safe strategies do not have the corresponding positions in the pay-off matrices so there is no stable solution. Alternatively, −3 + −2 ≠ 0 and so, again, there is no stable solution.

Sample questions and model answers *(continued)*

The values need to be made
positive by adding a fixed
value to each one.

(b) Adding 5 to each of the figures in Ashley's pay-off matrix gives:

$$
\begin{matrix}
10 & 2 & 8 \\
9 & 6 & 3 \\
2 & 11 & 1
\end{matrix}
$$

Using a, b and c to represent the probability that Ashley uses strategy A, B or C respectively, the linear programming problem is formulated as:

Maximise P subject to $a + b + c \leqslant 1$ where $a \geqslant 0$, $b \geqslant 0$, $c \geqslant 0$ and

$$P - 10a - 9b - 2c \leqslant 0$$
$$P - 2a - 6b - 11c \leqslant 0$$
$$P - 8a - 3b - c \leqslant 0$$

and P represents the value of the game.

2

The diagram shows an activity network for a project with the times given in days.

(a) Find the earliest start time and latest finish time for each activity.

(b) Find the critical path and give the time needed for completion of the project.

(a)

Use a forward scan for the
earliest start times.

Use a backward scan for
the latest finish times.

(b) The critical path is B, D, E, H and I. The time needed is 14 days.

Practice examination questions

1 A and B are the players in a two-person zero-sum game. Each may use one of two strategies when a game is played. The pay-off matrix for A is shown below.

	Player B	
Player A		Y	Z
W	4	8
X	6	3

(a) Find the best mixed strategy for each player.

(b) Find the value of the game.

2 Richard and Judy play a zero-sum game. The pay-off matrix for Richard is shown in the table.

		Judy		
		X	Y	Z
Richard	A	4	3	−4
	B	−3	6	8
	C	2	1	−6

(a) Find the play-safe strategy for each player.

(b) Show that there is no saddle point for the game.

(c) Explain why strategy C will not be part of Richard's mixed strategy.

(d) Find Richard's optimal mixed strategy.

3 A linear programming problem is formulated as:

Maximise $P = x + 2y + 3z$,

subject to $2x + 6y + z \leqslant 10$

$x + 4y + 5z \leqslant 16$

and $x \geqslant 0, y \geqslant 0, z \geqslant 0$.

(a) Set up an initial simplex tableau to represent the problem.

(b) Use the simplex algorithm to solve the problem.

State the corresponding values of x, y and z.

(c) Explain how you know that the optimum solution has been found.

Practice examination questions (continued)

4 An expedition is planned across a desert from S to T. The network shows the possible routes and distances in km between places where water and supplies are available. The optimal route is the minimax route which minimises the maximum distance between supply points.

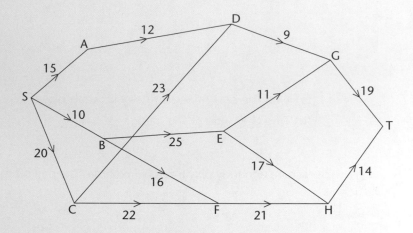

By completing the following table, or otherwise, use dynamic programming, working backwards from T, to find the minimax route and give the maximum distance between supply points on the route.

Stage	State	Action	Value	
1	G	GT	19	*
	H	HT	14	*
2	D	DG	Max(9, 19) = 19	
	E	EH		
	F	FH		
3	A	AD		
	B	BE		
		BF		
4				

Practice examination answers

Core 3 and Core 4 (Pure Mathematics)

1
$$\frac{1}{(1+x)^2} = (1+x)^{-2}$$
$$= 1 + (-2)x + \frac{(-2)(-3)}{2!}x^2 + \dots$$
$$= 1 - 2x + 3x^2 + \dots \quad |x| < 1$$
$$\sqrt{4+x} = \sqrt{4\left(1 + \frac{x}{4}\right)}$$
$$= 4^{\frac{1}{2}}\left(1 + \frac{x}{4}\right)^{\frac{1}{2}}$$
$$= 2\left(1 + \frac{x}{4}\right)^{\frac{1}{2}}$$
$$= 2\left(1 + (\tfrac{1}{2})\left(\frac{x}{4}\right) + \frac{(\tfrac{1}{2})(-\tfrac{1}{2})}{2!}\left(\frac{x}{4}\right)^2 + \dots\right)$$
$$= 2 + \frac{x}{4} - \frac{x^2}{64} + \dots \quad |x| < 4$$
$$\therefore \quad f(x) = 1 - 2x + 3x^2 + 2 + \frac{x}{4} - \frac{x^2}{64} + \dots$$
$$= 3 - 1\tfrac{3}{4}x + 2\tfrac{63}{64}x^2 + \dots \quad |x| < 1$$

Ignoring terms in x^3 and higher, $f(x) \approx a + bx + cx^2$, with $a = 3$, $b = -1\tfrac{3}{4}$, $c = 2\tfrac{63}{64}$.

The restriction is $|x| < 1$.

2 (a)
$$y = \cos^2 x \sin x$$
$$\frac{dy}{dx} = \cos^2 x(\cos x) + \sin x(-2\sin x \cos x)$$
$$= \cos^3 x - 2\cos x \sin^2 x$$
$$= \cos^3 x - 2\cos x(1 - \cos^2 x)$$
$$= \cos^3 x - 2\cos x + 2\cos^3 x$$
$$= 3\cos^3 x - 2\cos x$$

(b)
$$u = \cos x \Rightarrow \frac{du}{dx} = -\sin x$$
$$\frac{dx}{du} = -\frac{1}{\sin x} \Rightarrow \sin x \frac{dx}{du} = -1$$
$$\int \cos^2 x \sin x\, dx$$
$$= \int \cos^2 x \sin x \frac{dx}{du}\, du$$
$$= \int (-u^2)\, du$$
$$= -\tfrac{1}{3}u^3 + c$$
$$= -\tfrac{1}{3}\cos^3 x + c$$

(This can be also be done by direct recognition, see page 43.)

3
$$\frac{dy}{dx} = \frac{2x(x+1) - 1(x^2 - 4)}{(x+1)^2}$$
$$= \frac{x^2 + 2x + 4}{(x+1)^2}$$

When $x = 2$, $\dfrac{dy}{dx} = \dfrac{4}{3}$

$\therefore$ gradient of normal $= -\tfrac{3}{4}$

Equation of normal at $(2, 0)$:
$$y - 0 = -\tfrac{3}{4}(x - 2)$$
$$4y = -3x + 6$$
$$3x + 4y - 6 = 0$$

$\therefore$ $a = 3$, $b = 4$ and $c = -6$.

4 (a) $g(x) = (x+1)(x-1) = x^2 - 1$.

Since $x^2 \geqslant 0$ the range of g is $x \geqslant -1$.

(b)
$$fg(x) = f(x^2 - 1)$$
$$= 5(x^2 - 1) - 7$$
$$= 5x^2 - 12.$$

(c)
$$fg(x) = f(x)$$
$$\Rightarrow 5x^2 - 12 = 5x - 7$$
$$\Rightarrow 5x^2 - 5x - 5 = 0$$
$$\Rightarrow x^2 - x - 1 = 0$$
$$\Rightarrow x = \frac{1 \pm \sqrt{5}}{2}.$$

5 (a)
$$R\sin(x + a) \equiv 15\sin x + 8\cos x$$
$$\Rightarrow R\cos a = 15 \text{ and } R\sin a = 8 \text{ (a is acute)}$$
$$\Rightarrow R = \sqrt{15^2 + 8^2} = 17$$
$$\tan a = \tfrac{8}{15} \Rightarrow a = 28.07°$$
So $15\sin x + 8\cos x \equiv 17\sin(x + 28.07°)$.

(b) Max value is 17.

(c)
$$15\sin x + 8\cos x = 12$$
$$\Rightarrow 17\sin(x + 28.07°) = 12$$
$$\sin^{-1}(\tfrac{12}{17}) = 44.90°, \ 180° - 44.90° = 135.10°$$
$$x + 28.07 = 44.90° \Rightarrow x = 16.8°$$
$$x + 28.07 = 135.10° \Rightarrow x = 107.0°$$

Core 3 and Core 4 *(Pure Mathematics)* *(continued)*

6 $x\dfrac{dy}{dx} = y + yx \Rightarrow x\dfrac{dy}{dx} = y(1 + x)$

Separating the variables:
$$\dfrac{1}{y}\dfrac{dy}{dx} = \dfrac{1+x}{x}$$

$$\dfrac{1}{y}\dfrac{dy}{dx} = \dfrac{1}{x} + 1$$

$$\int \dfrac{1}{y}\,dy = \int \left(\dfrac{1}{x} + 1\right)dx$$

$\therefore \quad \ln y = \ln x + x + c$

When $x = 2$, $y = 4$,

$\Rightarrow \quad \ln 4 = \ln 2 + 2 + c$

$\Rightarrow \qquad c = \ln 2 - 2 \quad (\ln 4 = 2\ln 2)$

So $\quad \ln y = \ln x + x + \ln 2 - 2$

$$\ln\left(\dfrac{y}{2x}\right) = x - 2$$

$$\dfrac{y}{2x} = e^{x-2}$$

$$y = 2xe^{x-2}$$

7 (a) (i) $\quad y = (4x - 3)^9$

$\Rightarrow \dfrac{dy}{dx} = 9(4x - 3)^8 \times 4$

$\Rightarrow \dfrac{dy}{dx} = 36(4x - 3)^8.$

(ii) $\quad y = \ln(5 - 2x)$

$\dfrac{dy}{dx} = \dfrac{1}{5 - 2x} \times (-2)$

$= \dfrac{-2}{5 - 2x} \quad \left(\text{you can write this as } \dfrac{2}{2x - 5}\right).$

(b) (i) $\displaystyle\int (4x - 3)\,dx = \dfrac{1}{36}(4x - 3)^9 + C.$

(ii) $\displaystyle\int_3^4 \dfrac{1}{5 - 2x}\,dx = -\dfrac{1}{2}\int_3^4 \dfrac{-2}{5 - 2x}\,dx$

$= -\dfrac{1}{2}\left[\ln|5 - 2x|\right]_3^4$

$= -\dfrac{1}{2}\left(\ln|-3| - \ln|-1|\right)$

$= -\dfrac{1}{2}\ln 3.$

8 (a) $\displaystyle\int_0^1 xe^{-2x}\,dx = \left[x(-\tfrac{1}{2}e^{-2x})\right]_0^1 - \int_0^1 (-\tfrac{1}{2}e^{-2x})\,dx$

$= -\tfrac{1}{2}e^{-2} + \tfrac{1}{2}\displaystyle\int_0^1 e^{-2x}\,dx$

$= -\tfrac{1}{2}e^{-2} + \tfrac{1}{2}\left[(-\tfrac{1}{2}e^{-2x})\right]_0^1$

$= -\tfrac{1}{2}e^{-2} + \tfrac{1}{2}(-\tfrac{1}{2}e^{-2} - (-\tfrac{1}{2}))$

$= -\tfrac{3}{4}e^{-2} + \tfrac{1}{4}$

(b) (i) $\quad y = xe^{-x}$

$\dfrac{dy}{dx} = x(-e^{-x}) + e^{-x} = e^{-x}(1 - x)$

$\dfrac{dy}{dx} = 0$ when $x = 1$

When $x = 1$, $y = e^{-1} \Rightarrow A$ is point $(1, e^{-1})$.

(ii) $V = \pi\displaystyle\int_0^1 y^2\,dx = \pi\int_0^1 x^2 e^{-2x}\,dx$

$= \pi\left(\left[x^2(-\tfrac{1}{2}e^{-2x})\right]_0^1 - \int_0^1 -\tfrac{1}{2}e^{-2x}(2x)\,dx\right)$

$= \pi\left(-\tfrac{1}{2}e^{-2} + \int_0^1 xe^{-2x}\,dx\right)$

$= \pi(-\tfrac{1}{2}e^{-2} - \tfrac{3}{4}e^{-2} + \tfrac{1}{4})$ (using part (a))

$= \tfrac{\pi}{4}(1 - 5e^{-2})$

9 $\dfrac{dP}{dt} \propto P \Rightarrow \dfrac{dP}{dt} = kP$, where $k > 0$.

Separating the variables

$$\int \dfrac{1}{P}\,dP = \int k\,dt$$

$\Rightarrow \quad \ln P = kt + c$

$\Rightarrow \qquad P = e^{kt+c}$

$\Rightarrow \qquad P = Ae^{kt}$ (where $A = e^c$)

$t = 0$, $P = 500 \Rightarrow 500 = A$

$t = 10$, $P = 1000 \Rightarrow 1000 = 500e^{10k}$

$\therefore \qquad 2 = e^{10k}$

$\Rightarrow \quad 10k = \ln 2$

$\Rightarrow \qquad k = 0.1\ln 2$

$\therefore \quad P = 500e^{(0.1\ln 2)t}$

When $t = 20$,

$P = 500e^{0.1\ln 2 \times 20} = 500 \times 4 = 2000$

So the population is 2000 when $t = 20$.

Core 3 and Core 4 (Pure Mathematics) (continued)

10 (a)

$$x^3 + xy + y^2 = 7$$

$$3x^2 + x\frac{dy}{dx} + y + 2y\frac{dy}{dx} = 0$$

At (1, 2)

$$3 + \frac{dy}{dx} + 2 + 4\frac{dy}{dx} = 0$$

$$\Rightarrow 5\frac{dy}{dx} + 5 = 0 \text{ i.e. } \frac{dy}{dx} = -1$$

At (1, 2) gradient of normal = 1

Equation of normal is

$$y - 2 = 1(x - 1)$$

i.e. $y = x + 1$

(b) $x = t^2$, $y = t + \frac{1}{t}$.

$$\frac{dx}{dt} = 2t$$

$$\frac{dy}{dt} = 1 - \frac{1}{t^2} = \frac{t^2 - 1}{t^2}$$

$$\frac{dy}{dx} = \frac{dy}{dt} \times \frac{dt}{dx} = \frac{t^2 - 1}{2t^3}$$

$$\frac{dy}{dx} = 0 \text{ when } t^2 - 1 = 0 \Rightarrow t = \pm 1$$

When $t = 1$, coordinates are (1, 2).

When $t = -1$, coordinates are (1, −2).

The stationary points are at (1, 2) and (1, −2).

11 (a) $\sin 2x = 0$ when $2x = 0, \pi, 2\pi, \ldots$

i.e. when $x = 0, \frac{1}{2}\pi, \pi, \ldots$

$$\therefore a = \frac{1}{2}\pi$$

(b) $A = \int_0^{\frac{1}{2}\pi} y\,dx = \int_0^{\frac{1}{2}\pi} \sin 2x\,dx$

$$= -\frac{1}{2}\left[\cos 2x\right]_0^{\frac{1}{2}\pi}$$

$$= -\frac{1}{2}(-1 - 1)$$

$$= 1$$

Area of R is 1 square unit.

(c) $V = \pi \int_0^{\frac{1}{2}\pi} y^2\,dx$

$$= \pi \int_0^{\frac{1}{2}\pi} \sin^2 2x\,dx$$

$$= \frac{1}{2}\pi \int_0^{\frac{1}{2}\pi} (1 - \cos 4x)\,dx$$

$$= \frac{1}{2}\pi\left[x - \frac{1}{4}\sin 4x\right]_0^{\frac{1}{2}\pi}$$

$$= \frac{1}{2}\pi(\frac{1}{2}\pi - 0 - 0)$$

$$= \frac{1}{4}\pi^2$$

The volume generated is $\frac{1}{4}\pi^2$ cubic units.

12 (a) Referring to origin (0, 0)

$$\mathbf{a} = \begin{bmatrix} 2 \\ 2 \\ 0 \end{bmatrix} \text{ and } \mathbf{b} = \begin{bmatrix} 2 \\ 3 \\ 1 \end{bmatrix}$$

So $\mathbf{b} - \mathbf{a} = \begin{bmatrix} 0 \\ 1 \\ 1 \end{bmatrix}$

A vector equation for AB is

$\mathbf{r} = \mathbf{a} + \mu(\mathbf{b} - \mathbf{a})$ where μ is a scalar parameter

$$\therefore \mathbf{r} = \begin{bmatrix} 2 \\ 2 \\ 0 \end{bmatrix} + \mu\begin{bmatrix} 0 \\ 1 \\ 1 \end{bmatrix}$$

(b) If lines intersect, there are unique values for μ and λ for which

$$\begin{bmatrix} 2 \\ 2 \\ 0 \end{bmatrix} + \mu\begin{bmatrix} 0 \\ 1 \\ 1 \end{bmatrix} = \begin{bmatrix} 0 \\ 2 \\ 1 \end{bmatrix} + \lambda\begin{bmatrix} 1 \\ 0 \\ 1 \end{bmatrix}$$

Equating coefficients of $\mathbf{i}$, $\mathbf{j}$ and $\mathbf{k}$, in turn gives:

$2 = \lambda$

$2 + \mu = 2$

$\mu = 1 + \lambda$

There are no values of μ and λ that satisfy all three equations, so the lines have no common point.

(c) The angle between the lines is the angle between their direction vectors $\mathbf{d}_1 = \begin{bmatrix} 0 \\ 1 \\ 1 \end{bmatrix}$ and $\mathbf{d}_2 = \begin{bmatrix} 1 \\ 0 \\ 1 \end{bmatrix}$

$|\mathbf{d}_1| = \sqrt{1^2 + 1^2} = \sqrt{2}$ and $|\mathbf{d}_2| = \sqrt{1^2 + 1^2} = \sqrt{2}$

$$\mathbf{d}_1.\mathbf{d}_2 = \begin{bmatrix} 0 \\ 1 \\ 1 \end{bmatrix}.\begin{bmatrix} 1 \\ 0 \\ 1 \end{bmatrix} = 1$$

Let the angle be θ, then

$$\cos\theta = \frac{1}{\sqrt{2}\sqrt{2}} = 0.5$$

$$\Rightarrow \quad \theta = 60°$$

Statistics 2

1 If $\mu = 20$, $\bar{X} \sim N(20, \frac{16}{10})$,

i.e. $\bar{X} \sim N(20, 1.6)$.

Carrying out a two-tailed test at 5% level, reject H_0 if $z < -1.96$ where

$$z = \frac{17.2 - 20}{\sqrt{1.6}} = -2.213\ldots$$

Since $z < -1.96$, reject H_0.

There is evidence that μ is not 20.

2 (a) $f(x) = k$, $2 \leqslant x \leqslant 12$
Total probability = 1
$\Rightarrow 10 \times k = 1$, i.e. $k = \frac{1}{10}$
$\therefore \; f(x) = \frac{1}{10}$, $2 \leqslant x \leqslant 12$

(b) By symmetry, $E(X) = \frac{1}{2}(2 + 12) = 7$

$$E(X^2) = \frac{1}{10}\int_2^{12} x^2 \, dx = \frac{1}{10}\left[\frac{x^3}{3}\right]_2^{12} = 57\tfrac{1}{3}$$

$\text{Var}(X) = 57\tfrac{1}{3} - 7^2 = 8\tfrac{1}{3}$

or $\text{Var}(X) = \frac{1}{12}(12 - 2)^2 = 8.333\ldots$ (formula)

Standard deviation $= \sqrt{8\tfrac{1}{3}} = 2.887$ (3 d.p.)

$P(7 - 2.887 < X < 7 + 2.887)$
$\quad = P(4.113 < X < 9.887)$
$\quad = 0.1 \times (9.887 - 4.113)$
$\quad = 0.58$ (2 d.p.)

3 From the sample: $\bar{x} = \dfrac{\sum x}{n} = \dfrac{255.6}{9} = 28.4$

(a) $\bar{X} \sim N\left(\mu, \dfrac{\sigma^2}{n}\right)$, i.e. $\bar{X} \sim N\left(\mu, \dfrac{4^2}{9}\right)$.

90% confidence limits

$$= \bar{x} \pm 1.6449 \frac{\sigma}{\sqrt{n}}$$

$$= 28.4 \pm 1.6449 \times \frac{4}{\sqrt{9}}$$

$$= 28.4 \pm 2.1932$$

90% confidence interval

$$= (28.4 - 2.1932, \, 28.4 + 2.1932)$$

$$= (26.2, 30.6) \text{ (to 1 d.p.)}$$

(b) (i) $s = \sqrt{\dfrac{\sum(x - \bar{x})^2}{n - 1}}$

$$= \sqrt{\dfrac{113.22}{8}}$$

$$= 3.762 \text{ (3 d.p.)}$$

(ii) Since the sample size is small, the $t(8)$ distribution is considered. The critical t values, at the 95% level, are ± 2.306.

95% confidence limits

$$= 28.4 \pm 2.306 \times \frac{3.762}{\sqrt{9}} = 28.4 \pm 2.892$$

Interval $= (25.5, 31.3)$ (to 1 d.p.)

Statistics 2 *(continued)*

4 **(a)** $\displaystyle\int_{\text{all }x} f(x)\,\mathrm{d}x = 1$

so $\displaystyle 1 = k\int_0^1 (x - x^3)\,\mathrm{d}x$

$\displaystyle = k\left[\frac{x^2}{2} - \frac{x^4}{4}\right]_0^1$

$\displaystyle = k\left(\frac{1}{2} - \frac{1}{4}\right)$

$\displaystyle = \frac{1}{4}k$

$\therefore \quad k = 4$

(b) $\displaystyle \mu = E(X) = \int_0^1 xf(x)\,\mathrm{d}x = 4\int_0^1 x(x - x^3)\,\mathrm{d}x$

$\displaystyle = 4\int_0^1 (x^2 - x^4)\,\mathrm{d}x$

$\displaystyle = 4\left[\frac{x^3}{3} - \frac{x^5}{5}\right]_0^1$

$\displaystyle = 4\left(\frac{1}{3} - \frac{1}{5}\right)$

$\displaystyle = \frac{8}{15}$

(c) $\displaystyle P(X < 0.5) = 4\int_0^{0.5} (x - x^3)\,\mathrm{d}x$

$\displaystyle = 4\left[\frac{x^2}{2} - \frac{x^4}{4}\right]_0^{0.5}$

$= 0.4375$

(d) $\displaystyle E(X^2) = \int_0^1 x^2 f(x)\,\mathrm{d}x = 4\int_0^1 x^2(x - x^3)\,\mathrm{d}x$

$\displaystyle = 4\int_0^1 (x^3 - x^5)\,\mathrm{d}x$

$\displaystyle = 4\left[\frac{x^4}{4} - \frac{x^6}{6}\right]_0^1$

$\displaystyle = \frac{1}{3}$

$\mathrm{Var}(X) = E(X^2) - \mu^2$

$\displaystyle = \frac{1}{3} - \left(\frac{8}{15}\right)^2$

$= 0.0489 \ (3\ \text{s.f.})$

5 **(a)** $F(2) = 1$

$\Rightarrow c(8 - 4 - 1) = 1 \text{ so } c = \frac{1}{3}$

(b) $P(X > 1\frac{1}{2}) = 1 - F(1\frac{1}{2})$

$\displaystyle = 1 - \frac{1}{3}\left(4 \times 1\frac{1}{2} - (1\frac{1}{2})^2 - 1\right)$

$\displaystyle = \frac{1}{12}$

(c) For $0 \leqslant x \leqslant 1$,

$\displaystyle f(x) = \frac{\mathrm{d}}{\mathrm{d}x}(\tfrac{2}{3}x) = \tfrac{2}{3}$

For $1 \leqslant x \leqslant 2$,

$\displaystyle f(x) = \frac{\mathrm{d}}{\mathrm{d}x}\left(\tfrac{1}{3}(4x - x^2 - 1)\right)$

$\displaystyle = \tfrac{1}{3}(4 - 2x) = \tfrac{2}{3}(2 - x)$

For $x \leqslant 0$, $x \geqslant 2$, $f(x) = 0$

(d) $\displaystyle E(X) = \int_0^1 \frac{2}{3}x\,\mathrm{d}x + \int_1^2 \frac{2}{3}(2x - x^2)\,\mathrm{d}x$

$\displaystyle = \left[\frac{1}{3}x^2\right]_0^1 + \frac{2}{3}\left[x^2 - \frac{x^3}{3}\right]_1^2$

$\displaystyle = \frac{1}{3} + \frac{2}{3}\left(4 - \frac{8}{3} - (1 - \frac{1}{3})\right)$

$\displaystyle = \frac{7}{9}$

Statistics 2 *(continued)*

6 H_0: Left- or right-handedness and the ability to complete the task are independent.
H_1: There is an association between them.

Expected frequencies

Right-handed, completed task: $\dfrac{150 \times 120}{200} = 90$

	Completed	Not	Totals
Right	90	60	150
Left	30	20	50
Totals	120	80	200

$v = (2-1)(2-1) = 1$. Consider the $\chi^2(1)$ distribution. From tables, the critical (5%) value is 3.841, so H_0 is rejected if $\chi^2 > 3.841$.

O	E	$\sum \dfrac{(\lvert O - E \rvert - 0.5)^2}{E}$
83	90	0.469 ...
67	60	0.704 ...
37	30	1.408 ...
13	20	2.112 ...
$\sum O = 160$	$\sum E = 160$	4.694 ...

$$\chi^2 = \sum \frac{(\lvert O - E \rvert - 0.5)^2}{E} = 4.694 \text{ (3 d.p.)}$$

Since $\chi^2 > 3.841$, H_0 is rejected. There is evidence of an association between right- or left-handedness and the ability to complete the task in the given time.

7 X is the number of requests for the newspaper on a Monday, where $X \sim \text{Po}(15)$.

(a) $P(X < 16) = P(X \leqslant 15) = 0.5681$

(b) Let the number in stock be k.

We require $P(X > k) < 0.05 \Rightarrow P(X \leqslant k) \geqslant 0.95$

From tables, $P(X \leqslant 21) = 0.9469 < 0.95$

$P(X \leqslant 22) = 0.9673 > 0.95$

So there should be 22 copies in stock to meet the demand.

Mechanics 2

1 (a) $x = t^3 - 6t^2 + 3$
$\dot{x} = 3t^2 - 12t = 3t(t-4)$
When $t = 0$, $\dot{x} = 0$ so initially at rest. At rest again when $t = 4$
$4^3 - 6 \times 4^2 + 3 = -29$
Max distance in negative direction is 29 m.

(b) The particle is at its initial position when
$t^3 - 6t^2 = 0 \Rightarrow t^2(t-6) = 0$
Particle passes through initial position when $t = 6$.
This gives $\dot{x} = 3 \times 6^2 - 12 \times 6 = 36$ ms⁻¹.

2 (a) $\mathbf{r} = (4t^2 + 5)\mathbf{i} + 2t^3\mathbf{j}$
$\dot{\mathbf{r}} = 8t\mathbf{i} + 6t^2\mathbf{j}$

When $t = 1$, $\dot{\mathbf{r}} = 8\mathbf{i} + 6\mathbf{j}$ so the speed = 10 ms⁻¹.

(b) P is moving parallel to
$\mathbf{i} + 3\mathbf{j} \Rightarrow 6t^2 = 24t \Rightarrow 6t(t-4) = 0$
Since $t > 0$ this gives $t = 4$.

3 (a) 3 cm (by symmetry).

(b) $\bar{y} = \dfrac{24 \times 2 - 4 \times 3}{20} = 1.8$ cm

(c)

$\theta = \tan^{-1}\left(\dfrac{3}{1.8}\right) = 59.0^0$

4 (a) Area rectangle : area triangle = 3 : 1

so   mass of rectangle = 0.6 kg
mass of triangle = 0.2 kg

$\bar{x} = \dfrac{0.6 \times 7.5 + 0.2 \times 18\frac{1}{3}}{0.8} = 10.2$ cm

(b)

15 cm

R_A ←10.2 cm→ R_B
8 N

$15R_B = 10.2 \times 8 \Rightarrow R_B = 5.44$ N
$R_A = 8 - 5.44 = 2.56$ N

5

40 m

Vertically: $R = mg$

Horizontally: $F = \dfrac{m \times 15^2}{40}$

$\mu = \dfrac{F}{R} = \dfrac{15^2}{40 \times 9.8} = 0.57$

6

(a) Magnitude of acceleration $= \omega^2 r$
$= 3^2 \times 0.3$
$= 2.7$ ms⁻²

(b) Using $F = ma$: $T \cos\theta = 2 \times 2.7$
so: $T \times \frac{3}{5} = 5.4$
giving: $T = 9$ N

(c)

2×9.8 N

Vertically: $R + 9\sin\theta = 2 \times 9.8$
$R = 2 \times 9.8 - 9 \times 0.8 = 12.4$
$R = 12.4$ N

7 (a) The tension is perpendicular to the direction of motion.

(b) The acceleration is not constant.

(c) Maximum speed occurs at lowest point.

Loss of PE = gain in KE
$mgl = \frac{1}{2}mv^2$
so, maximum speed is given by $v = \sqrt{2gl}$

Mechanics 2 *(continued)*

8 (a)

Resolving parallel to the plane:

$F = 900 \times 9.8 \times \frac{1}{15} + 800$

$\quad = 1388$

The driving force is 1388 N.

(b) $P = Fv = 1388 \times 20 = 27760$

Power $\qquad = 27.76$ kW

(c) Using $F = ma$:

gives: $1388 - 800 = 900a \Rightarrow a = 0.65 \, ms^{-2}$.

9 (a) The velocity at time t is given by:

$\dot{x} = 12t^2 - 6t = 6t(2t - 1)$

(b) $\dot{x} = 0$ when $t = 0$ or $t = 0.5$ so the particle changes direction when $t = 0.5$

$t = 0 \Rightarrow x = 7$

$t = 0.5 \Rightarrow x = 6.75$

The distance travelled is 0.25 m.

(c) P is at its starting point when

$4t^3 - 3t^2 = 0$

$\Rightarrow t^2(4t - 3) = 0$

$\Rightarrow t = 0$, or $t = 0.75$

P returns to its start point after 0.75 s

10 (a) Using $F = ma$: $-kv^2 = m \dfrac{dv}{dt}$

$\dfrac{dv}{dt} = -3$ when $v = 20$

so: $-\dfrac{k}{m} = \dfrac{1}{400} \times -3$

$\Rightarrow \dfrac{dv}{dt} = -\dfrac{3}{400} v^2$

$\Rightarrow \displaystyle\int v^{-2} dv = \int -\dfrac{3}{400} \, dt$

giving: $-\dfrac{1}{v} = -\dfrac{3t}{400} + C$

when $t = 0$, $v = 20$

so: $-\frac{1}{20} = C$

This gives $\dfrac{1}{v} = \dfrac{3t}{400} + \dfrac{1}{20} = \dfrac{3t + 20}{400}$

$\Rightarrow v = \dfrac{400}{3t + 20}$ as required.

(b) When $t = 0$: KE $= \frac{1}{2} \times 2 \times 20^2 = 400$ J

When $t = 4$: KE $= \frac{1}{2} \times 2 \times 12.5^2 = 156.25$ J

Work done by retarding force has magnitude $400 - 156.25 = 243.75$ J.

Mechanics 2 (continued)

11 (a) At the lowest point, the loss of gravitational potential energy has been converted to elastic potential energy.

Using $mgh = \dfrac{\lambda x^2}{2l}$

gives $mg(2 + x) = \dfrac{5mgx^2}{4}$

$\Rightarrow 5x^2 - 4x - 8 = 0.$

$\Rightarrow x = 1.73$ or $x = -0.927$ to 3 s.f.

When the particle is brought to rest, the extension is 1.73 m so the object is 3.73 m below O.

(b) Using $T = \dfrac{\lambda}{l}x$

gives $T = \dfrac{5mg}{2} \times 1.73$

Using $F = ma$ gives:

$\dfrac{5mg}{2} \times 1.73 - mg = ma$

Taking $g = 9.8$ gives:

$a = 2.5 \times 9.8 \times 1.73 - 9.8$

$= 32.585$

The object has acceleration 32.6 ms^{-2} to 3 s.f. at its lowest point.

12 (a) Horizontally:

$T \sin 30° = \dfrac{0.2 \times 1.2^2}{0.4}$

$\Rightarrow T = 1.44$

The tension is 1.44 N.

Vertically:

$1.44 \cos 30° + R = 0.2 \times 9.8$

$\Rightarrow R = 0.713$

The reaction is 0.713 N to 3 s.f.

(b) Vertically: the new tension is given by:

$T \cos 30° = 0.2 \times 9.8 \Rightarrow T = 2.263$

Horizontally: $2.263 \sin 30° = 0.2\omega^2 \times 0.4$

giving: $\omega = 3.76$

The minimum angular speed is 3.76 rad sec^{-1}.

13 (a) By the conservation of energy principle:

KE gained = PE lost

giving: $\frac{1}{2} \times 0.1 \times v^2 = 0.1g \times 0.6(1 - \cos\theta)$

$\Rightarrow v^2 = 1.2g(1 - \cos\theta)$

(b) Resolving along BC:

$0.1g \cos\theta - R = \dfrac{0.1 \times 1.2g(1 - \cos\theta)}{0.6}$

rearranging gives:

$R = g(0.3 \cos\theta - 0.2)$

(c) $R = 0 \Rightarrow 0.3 \cos\theta = 0.2$

$\Rightarrow \cos\theta = \frac{2}{3}$

$\Rightarrow \theta = 48.2°$

Decision Mathematics 2

1 Assume A uses Strategy W with probability p and strategy X with probability $(1 - p)$.

Expected gain for A is:

$4p + 6(1 - p)$ if B uses strategy Y.

$8p + 3(1 - p)$ if B uses strategy Z.

Optimal when

$4p + 6(1 - p) = 8p + 3(1 - p)$

$\Rightarrow -2p + 6 = 5p + 3$

$\Rightarrow p = \frac{3}{7}$

Assume B uses strategy Y with probability q and strategy Z with probability $(1 - q)$.

Expected loss for B is:

$4q + 8(1 - q)$ if A uses strategy W

$6q + 3(1 - q)$ if A uses strategy X.

Optimal when

$4q + 8(1 - q) = 6q + 3(1 - q)$

$\Rightarrow -4q + 8 = 3q + 3$

$\Rightarrow q = \frac{5}{7}$

The optimal strategy for A is to use strategy W with probability $\frac{3}{7}$ and strategy X with probability $\frac{4}{7}$.

The optimal strategy for B is to use strategy Y with probability $\frac{5}{7}$ and strategy Z with probability $\frac{2}{7}$.

The value of the game is $5\frac{1}{7}$.

2 (a) The minimum row values are -4, -3 and -6. The maximum of these is -3 showing that the play-safe strategy for Richard is strategy B.

The minimum of the negatives of the column values are -4, -6 and -8. The maximum of these is -4 showing that the play-safe strategy for Judy is strategy X.

(b) $-3 + (-4) \neq 0$ so there is no saddle point.

(c) Comparing the row values shows that Richard's pay-offs are *always* less with strategy C than with strategy A. It follows that strategy C will not be a part of his mixed strategy.

(d) Assume Richard uses strategy A with probability p and strategy B with probability $(1 - p)$.

The expected gains for Richard based on the strategy used by Judy are:

X: $4p - 3(1 - p) = 7p - 3$

Y: $3p + 6(1 - p) = -3p + 6$

Z: $-4p + 8(1 - p) = -12p + 8$

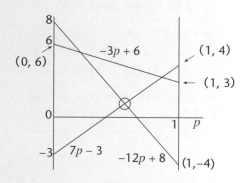

Optimal when $7p - 3 = -12p + 8$

$\Rightarrow p = \frac{11}{19}$

Richard's optimal strategy is to use strategy A with probability $\frac{11}{19}$ and strategy B with probability $\frac{8}{19}$.

Decision Mathematics 2 *(continued)*

3 (a) The initial tableau is:

P	x	y	z	s	t	val
1	−1	−2	−3	0	0	0
0	2	6	1	1	0	10
0	1	4	5	0	1	16

(b) The first iteration produces:

P	x	y	z	s	t	val
1	$-\frac{2}{5}$	$\frac{2}{5}$	0	0	$\frac{3}{5}$	$9\frac{3}{5}$
0	$\frac{9}{5}$	$\frac{26}{5}$	0	1	$-\frac{1}{5}$	$6\frac{4}{5}$
0	$\frac{1}{5}$	$\frac{4}{5}$	1	0	$\frac{1}{5}$	$3\frac{1}{5}$

The second iteration produces:

P	x	y	z	s	t	val
1	0	$\frac{14}{9}$	0	$\frac{2}{9}$	$\frac{5}{9}$	$11\frac{1}{9}$
0	1	$\frac{26}{9}$	0	$\frac{5}{9}$	$-\frac{1}{9}$	$3\frac{7}{9}$
0	0	$\frac{2}{9}$	1	$-\frac{1}{9}$	$\frac{2}{9}$	$2\frac{4}{9}$

The maximum value of P is $11\frac{1}{9}$.

This occurs when $x = 3\frac{7}{9}$, $y = 0$, $z = 2\frac{4}{9}$.

(c) This represents the optimum solution because there are no negative values in the objective row of the tableau.

4

Stage	State	Action	Value	
1	G	GT	19	*
	H	HT	14	*
2	D	DG	Max(9, 19) = 19	
	E	EH	Max(17, 14) = 17	*
	F	FH	Max(21,14) = 21	
3	A	AD	Max(12, 19) = 19	*
	B	BE	Max(25, 17) = 25	
		BF	Max(16, 21) = 21	*
	C	CD	Max(23, 19) = 23	
		CF	Max(22, 21) = 22	*
4	S	SA	Max(15, 19) = 19	*
		SB	Max(10, 21) = 21	
		SC	Max(20, 22) = 22	

The optimal minimax route is SADGT.

The maximum distance between supply points on the route is 19 km.

Notes

Index